Sydenham Edwards, William Darton

Plates and descriptions of such plants as grow wild in the environs

of London

Sydenham Edwards, William Darton

Plates and descriptions of such plants as grow wild in the environs of London

ISBN/EAN: 9783742888648

Manufactured in Europe, USA, Canada, Australia, Japa

Cover: Foto ©berggeist007 / pixelio.de

Manufactured and distributed by brebook publishing software
(www.brebook.com)

Sydenham Edwards, William Darton

Plates and descriptions of such plants as grow wild in the environs of London

I N D E X I.

In which the Plants contained in the Fourth Fasciculus are arranged according to the System of L I N N Æ U S.

Hippuris vulgaris

HIPPURIS VULGARIS. MARES-TAIL.

HIPPURIS *Lin. Gen. Pl.* MONANDRIA MONOGYNIA,

Cal. o. *Petala* o. *Stigma* fimpl.x. *Sem.* 1.

Raii Syn. Gen. 5. HERBÆ FLORE IMPERFECTO SEU STAMINEO VEL APETALO POTIUS.

HIPPURIS *vulgaris. Lin. Syft. Vegetab.* p. 51. *Sp. Pl.* p. 6. *Fl. Suec.* n. 2.

PINASTELLA. *Dillen. Nov. Gen.* p. 168.

LIMNOPEUCE. *Haller. Hift.* p. 264. *Vaillant. Mem. de l'Acad. anno* 1716, t. 1. f. 3.

POLYGONUM fæmina. *Matth. in Diofc.* p. 93a. *Dodon. Pempt.* p. 113.

EQUISETUM paluftre brevioribus foliis polyfpermon. *C. B. pin.* 15.

EQUISETUM paluftre alterum brevioribus fetis. *Park.* 1200.

CAUDA EQUINA fæmina. *Ger. emac.* 1114. *Raii Syn.* p. 136. *Hudfon. Fl. Angl. ed.* 2. p. 2. *Lightfoot Fl. Scot.* p. 70.

RADIX perennis, repens, geniculata, alba, geniculis plurimis fibris capillata.

CAULES plurimi, fefquipedales et ultra, erecti, fimplices, glabri, ftriati, teretes, fpongioti, *fig.* 1. medulla filiformi, compacta, in radicibus tenaci.

FOLIA verticillata, octo circiter, brevia, linearia, glaberrima, avenia, ad lentem punctata, punctis excavatis.

FLORES hermaphroditi plerumque, prefertim vere, ad finem æftatis plures fiemineos obfervavi, axillares, feffiles.

CALYX nullus.
COROLLA nulla.
STAMEN: FILAMENTUM unicum, apici germinis infidens, primo breviffimum, demiffo polline longitudine piftilli. ANTHERA biloba, purpurafcens, majufcula, *fig.* 2, 3.

PISTILLUM: GERMEN oblongum. STYLUS breviffimus, nudus. STIGMA fubulatum, album, ad lentem villofum, *fig.* 4, 5, 6.
PERICARPIUM nullum.
SEMEN unicum, oblongum, nudum, fuboffeum, intus album, medio fulcum, membranâ feu arillo tenui obtectum, *fig.* 7, 8, 9.

ROOT perennial, creeping, jointed and white, the joints furnifhed with numerous capillary fibres.

STALKS numerous, a foot and a half or more in height, upright, fimple, fmooth, ftriated, round, fpongy, *fig.* 1. the pith like a thread in the center, compact, and in the roots tough.

LEAVES growing in whirls, about eight in number fhort, linear, perfectly fmooth, without veins, dotted when magnified, the dots appearing hollow.

FLOWERS for the moft part hermaphrodite, efpecially in the fpring, at the clofe of the fummer I have obferved many of them to be female, growing in the ale af the leaves, and feffile.

CALYX none.
COROLLA none.
STAMEN: a fingle FILAMENT, fitting on the top of the germen, at firft very fhort, on fhedding the pollen becoming as long as the piftillum. ANTHERA compofed of two lobes, purplifh and rather large, *fig.* 2, 3.

PISTILLUM: GERMEN oblong. STYLE very fhort, naked. STIGMA tapering to a point, white and downy when magnified, *fig.* 4, 5, 6.
SEED-VESSEL none.
SEED fingle, oblong, naked, hard, white within, and in the center brown, covered with a thin membrane or arillus, *fig.* 7, 8, 9.

Greater fimplicity in the conftruction of a flower can fcarcely exift than in the *Hippuris*. Here we have neither calyx, corolla, nor feed-veffel; and thofe parts which are univerfally confidered as effential to the fructification are in the prefent inftance as few as poffible, there being only one ftamen, with its correfponding piftillum, yet perfect feed, and that in confiderable quantity, is produced.

The *Hippuris* here defcribed, which takes its name from the Greek 'Ιππυρις, five *Cauda equina*, is not the *Hippuris* of the firft Botanifts. They applied the term to our *Equifetum*, the *Hippuris* of LINNÆUS is the *Polygonum fæmina* of DIOSCORIDES, and arranged by his commentator MATTHIOLUS with our *Polygonum aviculare* and *Herniaria*. Succeeding Botanifts imagining, from the growth of its leaves, or from its producing feed, that it had better pretenfions to be ranked with the *Equifetum*, abfurdly enough called it *Cauda equina fæmina*, to which Mr. HUDSON could not well avoid giving the Englifh name of *Mares-tail.*

Although common in many parts of *Great Britain*, this plant is very rare about *London*, Mr. HUDSON mentions it as growing in a part of the New River near *Hornfey*, where it may ftill be found.

It flowers and produces its feeds from *June* to *Auguft.*

In running ftreams it is frequently extended to a great length; and we have been informed, that in fome rivers it is an exceedingly troublefome weed, which we can the more readily believe, having experienced its roots to be of the moft powerfully creeping kind.

A tranfverfe fection of its ftalk is a beautiful microfcopic object.

On examining this plant we have fometimes found its flowers to be female only.

Veronica montana

VERONICA MONTANA. MOUNTAIN SPEEDWELL.

VERONICA *Lin. Gen. Pl.* DIANDRIA MONOGYNIA.

Cor. Limbo 4 partito, lacinià infimà angustiore. *Capsula* bilocularis.

Raii Syn. Gen. 18. HERBÆ FRUCTU SICCO SINGULARI FLORE MONOPETALO.

VERONICA *montana* racemis lateralibus paucifloris, calycibus hirsutis, foliis ovatis rugosis crenatis petiolatis, caule debili. *Lin. Syst. Vegetab. Sp. Pl.* p. 36.

VERONICA caule procumbente, foliis hirsutis, cordatis, retusis, racemis paucifloris. *Haller. hist.* n. 539.

CHAMÆDRYI spuriæ affinis rotundifolia scutellata. *Bauh. pin.* 249.

ALYSSON Dioscoridis montanum. *Col. Ecph.* 1. 286.

VERONICA Chamædryoides, foliis pediculis oblongis insidentibus. *Raii Syn.* p. 281. Wild Germander with Leaves standing on long Foot-stalks.

Lightfoot Fl. Scot. p. 6.

Hudson. Fl. Angl. ed. 2. p. 74.

RADIX perennis, fibrosa, fibrillis prælongis, fuscis.

CAULES procumbentes, versus basin sæpe radicantes, teretes, pilis mollibus undique hirsuti, purpurascentes.

FOLIA opposita, petiolata, ovato-cordata, obtusiuscula, inæqualiter ferrata, hirsutula, nitidula, subtus purpurascentia, parum concava et bullata.

PETIOLI longitudine fere foliorem, hirsutissimi.

RACEMI laterales, alterni, subinde oppositi, tenues, hirsuti, paucislori.

PEDUNCULI alterni, hirsuti, bracteà lanceolatà suffulti.

CALYX: PERIANTHIUM tetraphyllum, soliolis subæqualibus, ovatis, basi angustatis, hirsutis, pilis ad lentem globuligeris. *fig.* 1.

COROLLA monopetala, rotata, ex purpureo cærulescens, supremà lacinià saturatius coloratà, una cum lateralibus venis cæruleis pictà, insimà minore immaculata, *tubus* brevissimus, albus. *fig.* 2.

STAMINA: FILAMENTA duo, tubo corollæ inserta, basi albida, curvata, medio crassiora; ANTHERÆ cæruleæ; POLLEN album. *fig.* 3.

PISTILLUM: GERMEN obovatum, hirsutum; STYLUS superne sensim incrassatus; STIGMA capitatum, album. *fig.* 4. 5.

PERICARPIUM: CAPSULA magna, orbiculata, emarginata, compressa. *fig.* 6.

SEMINA pauca, ovata, plana, flavescentia. *fig.* 7.

ROOT perennial and fibrous, the fibres very long and brown.

STALKS procumbent, often taking root towards the base, round, covered with soft hairs, and purplish.

LEAVES opposite, standing on footstalks, ovate cordate, a little blunt, unequally serrated, slightly hairy, somewhat shining, purplish underneath, a little hollow and cockled.

LEAF-STALKS almost the length of the leaves, and very hairy.

FLOWER-BRANCHES lateral, alternate, sometimes opposite, slender, hairy, supporting few flowers.

FLOWER-STALKS alternate, hairy, supported by a narrow floral-leaf.

CALYX: a PERIANTHIUM composed of four leaves, which are nearly equal, ovate, narrowed at the base, hairy, the hairs globular at the extremity when magnified. *fig.* 1.

COROLLA monopetalous, wheel-shaped, of a bluish purple colour, the uppermost segment more deeply coloured than the others, and together with the side ones streakt with blue veins, the lowermost least without any veins, the *tube* very short and white. *fig.* 2.

STAMINA: two FILAMENTS, inserted into the tube of the corolla, whitish at the base, bent, thickest in the middle; ANTHERÆ blue; POLLEN white. *fig.* 3.

PISTILLUM: GERMEN inversely ovate, hairy; STYLE towards the top gradually thickened; STIGMA forming a small white head. *fig.* 4. 5.

SEED-VESSEL: a large, round, flat CAPSULE nicked at top. *fig.* 6.

SEEDS few, ovate, flat and yellowish. *fig.* 7.

The *Veronica montana* is very similar in its general appearance to the *Chamædrys*, and of which, by some authors, it has been considered as only a variety; but this has arisen from a very superficial enquiry, as no two plants can be more distinct; LINNÆUS might indeed have selected a specific character which would effectually have removed every doubt of this kind, viz. the shape and size of the seed vessels, these in the *montana* are at least thrice as large as those of the *chamædrys*, they are also much r under and flatter, while the flower on the contrary are not more than half as large, and much less showy; when we have not these characters to assist us, the stalk and leaves will in general be sufficient, in the *chamædrys* the hairs grow on two sides of the stalk only, in the *montana* they grow all around it, in the *chamædrys* the leaves are generally sessile, in the *montana* they stand on footstalks.

These two plants differ also in their places of growth, the *montana*, so far as I have observed it, preferring moist and shady situations, whence the term *montana* seems ill-applied to it; near London, it is found plentifully in Chariton Wood, behind the Church, and flowers in June and July.

VALERIANA DIOICA. MARSH VALERIAN.

VALERIANA *Linnæi Gen. Plant.* TRIANDRIA MONOGYNIA.

Cal. o *Cor.* 1. petala, bafi hinc gibba, fupera. *Sem.* 1.

VALERIANA *dioica* floribus triandris dioicis foliis pinnatis integerrimis. *Lin. Syft. Vegetab. Sp. Pl. p. 44. Fl. Suec. n. 35.*

VALERIANA foliis radicalibus petiolatis ovatis ; caulinis pinnatis, fexu diftinĉta. *Haller. hift.* 208.

VALERIANA *dioica. Scopoli Fl. Carn. n.* 40.

VALERIANA paluftris minor. *Bauhin. p.* 164.

VALERIANA minor. *Ger. em.* 1075.

VALERIANA fylveftris minor. *Park.* 122.

Raii Syn. p. 200. Small wild Valerian, or Marfh Valerian. *Hudfon. Fl. Angl. ed.* 2. *p.* 12. *Lightfoot Fl. Scot. p.* 85.

RADIX perennis, geniculata, repens, craffitie pennæ coracis, allida, rubore aliquando tinĉta, odore fubaromatico valerianæ fylveftris.

ROOT perennial, jointed, creeping, the thicknefs of a crow-quill, white, fometimes tinged with red, having nearly the fame aromatic fmell as the wild valerian.

CAULIS pedalis aut fefquipedalis, erectus, fimplex, tetragonus, ftriatus, lævis ; rami pauci, ftriati.

STALK a foot, or a foot and a half high, upright, unbranched, four-cornered, ftriated and fmooth ; branches, few and ftriated.

FOLIA oppofita, radicalia integerrima, ovata, obtufa, caulina pauca, pinnatifida, pinnis duodecim circiter, venofis, obtufe ferratis.

LEAVES oppofite, the radical ones entire, ovate, obtufe, thofe of the ftalk few, pinnatifid, pinnæ about twelve in number, veiny, and obtufely ferrated.

FLORES fubcorymbofi, rubelli, dioici, femineis multo minoribus, *fig.* 1. flor. femin. magn. nat. *fig.* 2. flor. mafc.

FLOWERS forming a kind of corymbus, of a pink colour, and dioicous, the female flowers much the fmalleft. *fig.* 1. a female flower of its natural fize. *fig.* 2. a male flower.

BRACTEÆ plurimæ, lanceolatæ, floribus fubjectæ.

BRACTEÆ numerous, lanceolate, placed beneath the flowers.

Flos Femin.

CALYX vix ullus, margo fuperus. *fig.* 7.

COROLLA monopetala, tubus a latere inferiore gibbus, nectariferus ; *limbus* quinquefidus, laciniis obtufis, fubæqualibus ; antherarum rudimenta intra tubum cernantur.

PISTILLUM : GERMEN inferum, ovatum, compreffum, fulcatum, longitudine fere corollæ ; STYLUS albus, fuperne paulo incraffatus, corolla paulo longior, obliquus ; STIGMA trifidum. *fig.* 6, 7, 8, 9.

SEMEN ovato-oblongum, pallide fufcum, hinc carinatum, illinc trinerve, pappo pilofo coronatum. *fig.* 11, 12, 13.

Female Flower.

CALYX fcarce any, being only a prominent rim furrounding the top of the germen. *fig.* 7.

COROLLA monopetalous, the *tube* gibbous on the under fide, and containing honey ; the limb divided into five fegments, which are blunt and nearly equal ; rudiments of Antheræ are vifible within the tube. *fig.* 1.

PISTILLUM : GERMEN placed below the corolla, ovate, flat, grooved, nearly the length of the corolla ; STYLE white, fomewhat thickened near the top, a little longer than the corolla, oblique ; STIGMA trifid. *fig.* 6, 7, 8, 9.

SEED of an ovate oblong fhape, and pale brown colour, a fingle rib on one fide, and three on the other, crowned with a feathery down, *fig.* 11, 12, 13.

Flos Masc.

CALYX et corolla ficut in fem. *fig.* 2.

STAMINA : FILAMENTA tria, filiformia, corolla longiora ; ANTHERÆ albæ ; feu pallide rubentes, *fig.* 5. Piftillum imperfectum in centro floris. *fig.* 10.

Male Flower.

CALYX and corolla the fame as in the female. *fig.* 2.

STAMINA : three FILAMENTS filiform, longer than the corolla ; ANTHERÆ white, or pale red, *fig.* 5. an imperfect Piftillum in the center of each flower. *fig.* 10.

There are few plants in which nature fports more than in the Valerians, even out of the *four* fpecies which we have growing wild with us, *one* is monandrous, viz. the *rubra*, and *another* dioicous as the prefent. Thefe deficiencies in their claffical character are however the lefs to be lamented, as they furnifh excellent fpecific diftinctions.

The *dioica* is found only in wet and boggy fituations ; in the meadows and ofier-grounds about *Batterfea* it grows abundantly ; its bloffoms before they open are of a bright red colour, and being collected into fmall heads, are very confpicuous among the herbage in the month of April ; in June and July it produces its downy feeds, which, for their beauty and fingular manner of expanding their pappus or down, are highly deferving the attention of the curious.

The roots having a fimilar fmell, and probably the fame medicinal virtues, as the officinal Valerian, may be fubftituted in lieu thereof, if neceffary.

What Scopoli afferts of this plant is fo contrary to the common opinion of botanifts and our own obfervations, that we cannot forbear tranfcribing his own words ; they will either prove that his obfervations are not to be depended upon, or that this plant puts on a very different appearance in *Carniola* than it does in the other parts of *Europe*.

" Millena fpeciminis examinavi et nunquam vidi flores dioicos, fed nunc omnes hermaphroditos, nunc filamento
" uno aliove caftrato inftructos, nunc mafculos et femineos in eadem planta, ita tamen ut mafculi flores con-
" tingerent rudimentum germinis et ftyli ; fine feminibus vero perfectis nullam hactenus inveni plantam."

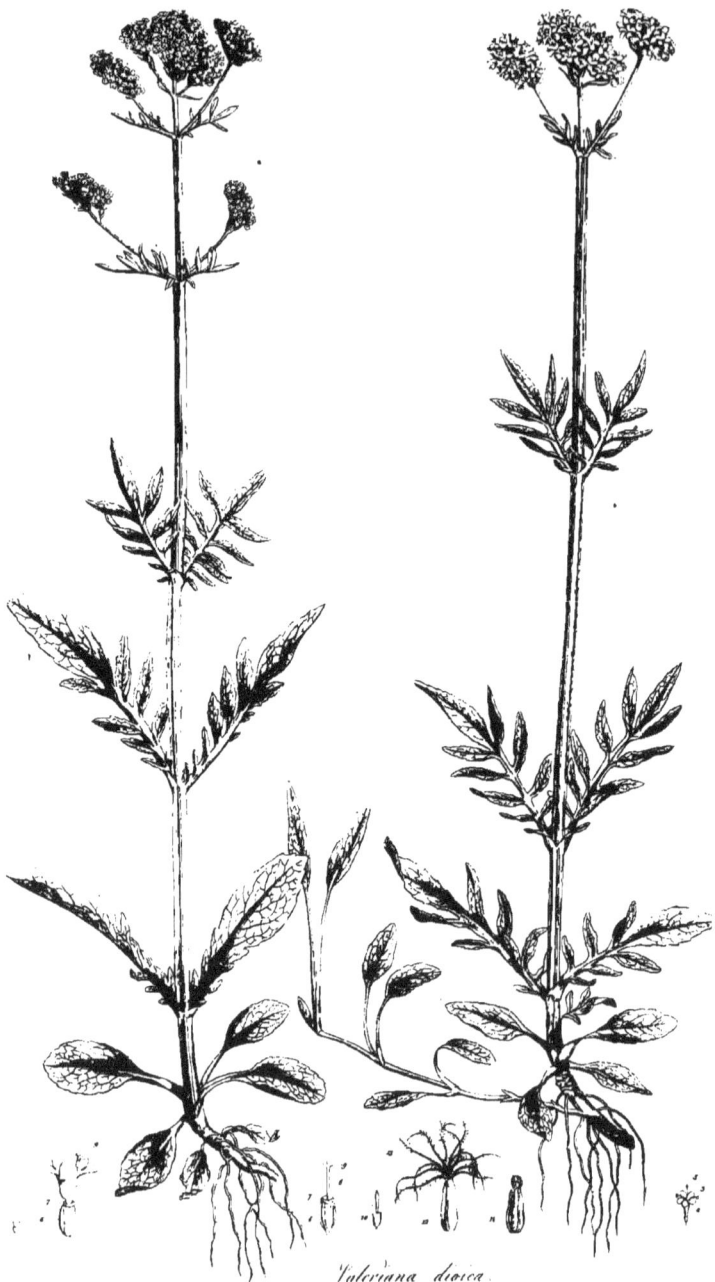

Valeriana dioica.

SCIRPUS MARITIMUS. ROUND-ROOTED OR SEA CLUB-RUSH.

SCIRPUS *Lin. Gen. Pl.* TRIANDRIA MONOGYNIA.

Glumæ paleaceæ, undique imbricatæ. *Cor.* o. *Sem.* 1 imberbe.

Raii Syn. Gen. 28. HERBÆ GRAMINIFOLIÆ FLORE NON CULMIFERÆ IMPERFECTO SEU STAMINEO.

SCIRPUS *maritimus* culmo triquetro, panicula conglobata foliacea spicularum squamis trifidis: intermedia subulata. *Lin. Syst. Vegetab.* p. 86. *Sp. Pl.* p. 74. *Fl. Suec.* n. 47.

SCIRPUS *maritimus. Scopoli Fl. Carn.* n. 57.

GRAMEN cyperoides panicula sparsa majus. *Baub. pin.* 6.

GRAMEN cyperoides paluftre panicula sparsa. *Parkins*, 1266. *Raii Syn.* p. 425. Water or Marsh Cyperus Grass, with a sparfed panicle.

CYPERUS rotundus littoreus inodorus. *Lob. Ic.* 77. rotundus inodorus Anglicus. *C. B. Pin.* 14. rotundus littoreus. *Ger. em.* 31. *Park.* 1264. *Raii Syn.* p. 426. Round-rooted Baftard Cyperus. *Hudfon. Fl. Anglic.* p. 21. *Lightfoot Fl. Scot.* p. 89.

RADIX perennis, repens, crafitie calami scriptorii, pallide fufca, ftolonibus fub finem anni apice bulbofis.

CULMUS bi feu tripedalis, erectus, foliofus, triqueter, angulis fubafperis.

FOLIA plurima, feptem five octo, pedalia aut fefquipedalia, lineas duas lata, fenfim acuminata, ad carinam et oras vix afpera, bafi vaginata, vagina minutiffime ftriata, nitida.

INVOLUCRUM: folia plerumque duo, rarius tria, inæqualia, longa, ad oras et carinam afpera.

PANICULA terminalis, maxime varia, aliquando enim conftat fpiculis quinque vel fex conglobatis feffilibus, faepius vero praeter has utrinque oritur pedunculus, tres quatuor vel quinque gerens fpiculas.

PEDUNCULI glabri, nudi ad fpiculas fubincraffati.

SPICULÆ magnæ, unciales fere, ovatæ; acutæ, primo atro purpureæ, demum ferrugineæ, fquamis undique imbricatæ.

CALYX: *Squamæ* fufcæ, corrugatæ, fcariofæ, carinatæ, apice faepius tridentatæ, dente medio fubulato, in infimis flofculis longiore. *fig.* 1. 2.

COROLLA nulla.

STAMINA: FILAMENTA tria, alba, latiufculæ. ANTHERÆ, flavae, llineares, membrana alba minuta terminatæ. *fig.* 3.

PISTILLUM: GERMEN obovatum, minimum, glabrum. STYLUS fubulatus, longitudine ftaminum. STIGMATA tria, capillaria. *fig.* 4.

VILLI quatuor aut quinque, ad bafin geminis, albi, erecti ad lentem retrorfum aculeati, geminfi longiores. *fig.* 5.

SEMEN unicum, fubtriquetrum, acuminatum, fufcum, nitidum. *fig.* 6.

ROOT perennial, creeping, the thicknefs of a goofequill, of a pale brown colour, the fhoots at the end of the year bulbous at their extremities.

STALK two or three feet high, upright, leafy, three-cornered, the angles fomewhat rough.

LEAVES numerous, feven or eight, a foot or a foot and a half in length, two lines in breadth, gradually tapering to a point, the keel and edges fcarcely rough, forming a fheath at bottom, which is ftriated and gloffy.

INVOLUCRUM confifts generally of two, rarely of three leaves, which are long, unequal and rough on the edges and keel

PANICLE terminal, and extremely various, fometimes it confifts of only five or fix cluftered fpiculæ, but for the moft part, befides thefe, a flower-ftalk arifes on each fide, bearing three, four, or five fpiculæ more.

FLOWER-STALKS fmooth, naked, fomewhat thickened at the fpiculæ.

SPICULÆ large, almoft an inch in length, ovate, pointed, at firft of a blackifh purple colour, afterwards ferruginous, covered with fcales on every fide.

CALYX: *Scales* brown, wrinkled, fonorous to the touch, keeled, having the tip generally furnifhed with three teeth, of which the middle one runs out to a long point, in the lowermoft flowers this is longeft. *fig.* 1. 2.

COROLLA wanting.

STAMINA: three FILAMENTS, white and broadifh. ANTHERÆ yellow, linear, tipt with a minute white membrane. *fig.* 3.

PISTILLUM: GERMEN inverfely ovate, very fmall and fmooth. STYLE tapering, the length of the ftamina. STIGMATA three, capillary. *fig.* 4.

HAIRS four or five at the bafe of the germen, white, upright, when magnified having prickles which crook backward, longer than the germen. *fig.* 5.

SEED fingle, fomewhat three-cornered, pointed, brown, and fhining. *fig.* 6.

LINNÆUS remarks, that this fpecies cloaths the fea-fhores as the Bulrufh does the borders of the inland lakes; but it is frequently found where the water is not falt, as in the river *Tbames*, and on the edges of the creeks running from it. In the *Ifle of Shepey* it fills almoft every ditch, and appears to be more perfectly at home.

It flowers from June to Auguft.

The older Botanifts made feveral fpecies of this plant, which LINNÆUS has very properly referred to varieties only. They did not attend to the œconomy of the plant, or they would have found, that the roots, in every variety, were bulbous at the extremities in the autumn, nor to the circumftances of fituation, &c. or they would have feen this plant fometimes fhorter, foractimes taller, fometimes with a fimple, fometimes with a branched panicle as is reprefented on the plate.

We know of no ufe to which this elegant fpecies of Club-rufh is applied. The roots have a remarkably fweet tafte, and probably are very nutritious.

Swine are extremely fond of the roots of the *Scirpus paluftris*, which the Swedifh peafants collect and fodder them with in the winter: the roots of the prefent fpecies, being much larger, would we conceive be much preferable for this or fimilar purpofes.

Scirpus maritimus.

PANICUM *Lin. Gen. Pl.* TRIANDRIA DIGYNIA.

Cal. 3-valvis : valvula tertia minima.

Raii Syn. Gen. 27. HERBÆ GRAMINIFOLIÆ FLORE IMPERFECTO CULMIFERÆ.

PANICUM *viride* fpica tereti, involucellis bifloris fafciculato-pilofis, feminibus nervofis. *Lin. Syft. Vegetab.* p. 502. *Sp. Pl.* p. 83.

PANICUM fpica unica, flofculis feffilibus folitariis, fetis numerofis. *Haller. Hift.* n. 1542.

GRAMEN paniceum fpica fimplici. *Baub. Pin.* 8.

GRAMEN panici effigie fpica fimplici. *Ger. emac.* 17.

GRAMEN paniceum fpica fimplici lævi. *Raii Syn.* p. 393. Panic-Grafs, with a fingle fmooth ear. *Hudfon Fl. Angl.* ed. 2. p. 24.

RADIX annua, fibrofa.

CULMI erecti, pedales et ultra, fimplices, fubinde ramofi, quatuor aut quinque geniculis diftincti, læves.

FOLIA palmaria et ultra, lineas duas, tres, quatuorve quandoque lata, acuminata, lævia, ad margines afpera, in apricis fæpe fanguinea; foliorum *Vagina* ftriata, lævis, ad internam foliorum bafin, loco membranulæ in pilos fubtiliffimos lineam dimidiam aut paulo plus longos terminata, qui pili etiam quandoque vaginæ margines fupremas ipfique foliorum bafi proximas occupant.

SPICA fimplex, teres, cylindracea, uncialis, fefquiuncialis et ultra, craffitie pennæ anferinæ majoris, aliquando tota fpadicea vel atro-purpurea, alias ex viridi lutefcens, luteis pilis, aliquando rubris donata, denfe coagmentatis conflans fpiculis, molliufcula, veftibus nequaquam adhærens; fetæ feu pili plurimi, erecti, tortuofi, flofculis triplo longiores, ad lentem aculeati, aculeis erectis. *fig.* 1. 2.

CALYX: *Gluma* uniflora, trivalvis, valvulis duabus oppofitis, æqualibus, ovatis, obtufis, nervofis, *fig.* 4. tertia minima, inferne pofita. *fig.* 3.

COROLLA: bivalvis, valvulæ ovatæ, concavæ, nitidæ, fubæquales. *fig.* 5.

STAMINA: FILAMENTA tria, capillaria, breviffima, corollam paulo excedentia. ANTHERÆ minimæ, purpureæ. *fig.* 6.

PISTILLUM: GERMEN ovatum. STYLI duo, capillares. STIGMATA plumofa, alba. *fig.* 7.

SEMEN unicum, fubovatum, tectum, hinc convexum, obfolete nervofum, inde plauiufculum. *fig.* 8.

ROOT annual, and fibrous.

STALKS upright, a foot high or more, fimple, now and then branched, furnifhed with four or five joints, and fmooth.

LEAVES about a hand's breadth or more in length, two or three lines, and fometimes more, in breadth, pointed, fmooth, rough on the edges, in open fituations often of a blood-red colour; *Sheath* of the leaves ftriated, fmooth, terminated at the inner bafe of the leaf, inftead of a membrane, by very fine hairs, about half a line or fomewhat more in length, which fometimes alfo occupy the edges of the fheath on its upper part, and of the leaves at their bafe.

SPIKE fimple, round, cylindrical, an inch, an inch and a half or more in length, the thicknefs of a large goofe quill, fometimes wholly of a reddifh purple colour, at others greenifh yellow, furnifhed with yellowifh, and fometimes reddifh hairs, compofed of fpiculæ clofely compacted, foft to the touch, never adhering to garments; fetæ or hairs numerous, upright, crooked, thrice the length of the flofcules, when magnified furnifhed with fmall prickles, which are upright. *fig.* 1. 2.

CALYX: a *Glume* of one flower, and three valves, two of which are oppofite, equal, ovate, obtufe, and ribbed, *fig.* 4. the third is very minute, and placed below the others. *fig.* 3.

COROLLA compofed of two valves, which are ovate, hollow, fhining, and nearly equal. *fig.* 5.

STAMINA: three capillary FILAMENTS, very fhort, a little longer than the corolla. ANTHERÆ very fmall, and purple. *fig.* 6.

PISTILLUM: GERMEN ovate. STYLES two, capillary. STIGMATA feathery and white. *fig.* 7.

SEED fingle, fomewhat ovate, covered, convex, and faintly ribbed on one fide, on the other flattifh. *fig.* 8.

In a former part of this work we gave figures of the *Panicum crufgalli* and *fanguinale*: we here prefent our readers with two more, being the whole of this genus growing near *London*.

The *viride* is with us the moft common of the four; yet at a diftance from town it appears to have few *habitats*. Mr. HUDSON particularizes *Martha's Chapel* near *Guildford*; *Batterfea Fields* is the only place where we find this, and the others, all of which flower about the fame period, viz. *Auguft* and *September*.

To correfpond with its name, the *viride* fhould be always of a green colour; but we often find its foliage red, and its fpikes reddifh-brown, and the *verticillatum* vice verfa: we are not therefore to look for an infallible guide in its colour, but the fpike will always diftinguifh it from the *verticillatum*. Between thefe two, indeed, there is a more fenfible difference to the touch than betwixt the *Alopecurus pratenfis* and *Phleum pratenfe*; the hairs in the fpike of the *viride* are much longer than thofe of the *verticillatum*, and though the microfcope difcovers them to be prickly, *vid. fg.* 1. 2. yet thefe being upright difcover no manifeft roughnefs.

Agriculturally it may be confidered rather as a weed than an ufeful grafs.

Sparrows are remarkably fond of its feeds: the whole of this genus, when cultivated in a garden, require to be protected from them.

Panicum viride.

Panicum verticillatum.

PANICUM. *Lin. Gen. Pl.* TRIANDRIA DIGYNIA.

Cal. trivalvis, valvula tertia minima.

Raii Syn. Gen. 27. HERBÆ GRAMINIFOLIÆ FLORE IMPERFECTO CULMIFERÆ.

PANICUM *verticillatum* fpica verticillata racemulis quaternis, involucellis unifloris bifetis, culmis diffufis. *Lin. Syſt. Vegetab.* p. 89. *Sp. Pl.* p. 82.

PANICUM fpica unica paniculata, fetis paucioribus. *Haller. Hiſt.* n. 1543.

GRAMEN paniceum fpica afpera. *Bauh. Pin.* p. 8.

PANICUM vulgare fpica fimplici et afpera. *Inſt.* 515. *Scheucb. Agroſt.* 47. *Raii Syn.* p. 394. Rough-eared Panic-Grafs. *Hudſon Fl. Angl. ed.* 2. p. 24.

Fig. 1. Racemula ramofa magn. nat.	*Fig.* 1. One of the fmall branched racemi of its natural fize.
Fig. 2. Pars ejufdem auct.	*Fig.* 2. A part of the fame magnified.
Fig. 3. Glumæ calycis auct.	*Fig.* 3. The glumes of the calyx magnified.
Fig. 4. Corolla.	*Fig.* 4. The Corolla.
Fig. 5. Stamina.	*Fig.* 5. The Stamina.
Fig. 6. Piſtillum.	*Fig.* 6. The Piſtillum.
Fig. 7. Semen magn. nat.	*Fig.* 7. The feed of its natural fize.
Fig. 8. Idem auct.	*Fig.* 8. The fame magnified.

The *Panicum verticillatum* in its general habit agrees exactly with the *viride*, but in the fize and form of the fpike, and the parts compofing it, it differs very materially : the whole plant is generally one-third larger than that of the viride ; the fpike is larger, and much lefs compact ; it is evidently compofed of little branches, which grow fomewhat in whirls, whence its name. The fetæ or hairs of the fpike are confiderably fhorter than thofe of the viride, and differ from them particularly in being hooked, fo that the fpike drawn over the back of the hand or cuff of the coat adheres very ſtrongly ; and where feveral fpikes grow near each other, they are very apt, from this caufe, to become entangled.

I found this plant growing laſt year very fparingly in the Gardeners Grounds *Baterſea Fields*, with the *viride*, and flowering at the fame time. Mr. RAY defcribes it as having been found in a Turnip Field betwixt *Putney* and *Roehampton*, alfo beyond the *Neat-houſes* by the *Thames* fide, going from the Horfe Ferry above *Weſtminſter* to *Chelſea*. SCHEUCHZER remarks, that it is a troublefome weed in the gardens at *Paris*.

Panicum sanguinale.

PANICUM SANGUINALE. COCK'S-FOOT PANIC-GRASS.

PANICUM *Lin. Gen. Pl.* TRIANDRIA DIGYNIA.
    Cal. 3-valvis: valvula tertia minima.
PANICUM *sanguinale* spicis digitatis basi interiore nodosis, flosculis geminis muticis, vaginis foliorum
    punctatis. *Lin. Syst. Vegetab.* p. 90. *Sp. Pl.* 84.
DIGITARIA foliis subhirsutis, caule debili, spicis verticillatis. *Haller Hist.* n. 1526.
DIGITARIA *sanguinalis. Scopoli Fl. Carn.* n. 72.
GRAMEN dactylon latiore folio. *Bauhin Pin.* 8.
ISCHÆMON sylvestre latiore folio. *Parkins.* 1178.
ISCHÆMON vulgare. *Ger. emac.* 27. Cock's-foot-grass. *Raii Syn.* p. 399. *Scheuch. Agrost.* 101.
    Schreb. Agrost. t. 16. Hudson Fl. Angl. ed. 2. p. 25.

RADIX annua, fibrosa.

CULMI ex una radice plures, spithamæi, pedales, subrubentes, subinde ramosi, adscendentes, infracti, tenues valde et debiles, quatuor plerumque geniculis distincti.

FOLIA uncialia, sesquiuncialia, et biuncialia, lineas duas aut duas cum dimidia lata, acuta, ad unum latus sæpe undulata, superne et inferne raris pilis hirsuta, marginibus ad lentem minutissime serrulatis; vagina striata, valde pilosa, pilis e punctis prominulis prodeuntibus.

SPICÆ ternæ, quaternæ, et quinæ plerumque, in cultis sæpe plures, digitatim summo culmo insidentes, sesquiunciam ad quatuor uncias longæ, filiformes, vel eodem loco oriundæ omnes, vel alternatim e summo culmo prodeuntes, et exiguo ab invicem spatio discretæ, purpureæ, aut ex purpureo et viridante mixtæ.

SPICULÆ secundæ, binæ, rachi adpressæ, pedicellatæ, pedicello altero longiore, longitudine spiculæ, ovato-lanceolatæ, acutæ, nunc purpurascentes, nunc virides. *fig.* 1, 2, 3, 4. *auct.*

CALYX trivalvis, persistens, infima minima, brevissima, nudo oculo vix conspicua, *fig.* 2. secunda et tertia oppositis, inæqualibus, acutis, nervosis, margine scabris, superiore longitudine corollæ quam margine suo amplectitur, inferiore dimidio breviore. *fig.* 3, 4.

COROLLA: bivalvis, valvulis æqualibus, glabris, alterâ alteram recipiente, *fig.* 8. ubi membranula ad basin earum pingitur, *fig.* 6, 7. disjunctæ apparent.

STAMINA: FILAMENTA tria, capillaria, corollâ paulo longiora; ANTHERÆ breves, parvæ, purpurascentes, utrinque bifurcæ. *fig.* 9.

PISTILLUM: GERMEN oblongum; STYLI duo, filiformes, longitudine staminum; STIGMATA plumosa, purpurea. *fig.* 10.

SEMEN minimum, oblongum glumis calycinis æque ac corollaccis inclusum, *fig.* 11, 12. denudatum, *fig.* 13. magnit. nat. *fig.* 14, 15. *auct.*

ROOT annual and fibrous.

STALKS several from one root, a span or a foot in height, of a reddish colour, sometimes branched, bending upward, crooked, very slender and weak, and generally furnished with four joints.

LEAVES an inch, an inch and a half, or two inches in length, and from two to two lines and a half in breadth, pointed, often waved on one side, on both sides beset with a few hairs, the edges when magnified very finely sawed; the sheath striated, very hairy, the hairs proceeding from little prominent points.

SPIKES sitting on the top of the stem, generally three, four or five together, often more in cultivated places, branching out like fingers, from one inch and a half to four inches in length, filiform, all of them proceeding from the same point, or growing alternately, leaving a small space betwixt them, of a dark purple colour, or purple and green mixed together.

SPICULÆ growing one way, pressed to the rachis, standing on foot-stalks, the longest of which is of the length of one of the spiculæ, ovato-lanceolate and pointed, sometimes purplish, and sometimes green. *fig.* 1, 2, 3, 4. *magnified.*

CALYX composed of three valves, and permanent, the lowermost very minute and short, scarcely perceptible by the naked eye, *fig.* 2. the second and third opposite, unequal, pointed, rib'd, rough on the edges, the upper one the length of the corolla, which it enfolds with its margin, the lower one half its length. *fig.* 3, 4.

COROLLA: composed of two valves which are equal and smooth, the one receiving the other, *fig.* 8. where a small membrane is painted at their base; at *fig.* 6. and 7. they appear disjoin'd.

STAMINA: three FILAMENTS, very slender, a little longer than the corolla; ANTHERÆ small, short, purplish, forked at each end. *fig.* 9.

PISTILLUM: GERMEN oblong; STYLES two, filiform, the length of the stamina; STIGMATA feather'd and purple. *fig.* 10.

SEED very small, oblong, inclosed by the glumes of the calyx, as well as of the corolla, *fig.* 11, 12. stripped of these, *fig.* 13. of its natural size, and magnified at *fig.* 14, 15.

Modern Botanists are divided in their opinions respecting the genus of this plant, LINNÆUS, SCHREBER, and others considering it as a *Panicum*; ADANSON, HEISTER, HALLER and SCOPOLI arranging it under a new genus, viz. *Digitaria*, a name adapted to the particular disposition of its spikes, but as that particular disposition has little to do with its generic character, and as it has a triphyllous calyx, which LINNÆUS, with much propriety, considers as a principal character of the *Panicum*, we follow him in preference to the others, however respectable. We may observe, that the exterior leaf forming this triphyllous calyx is so very minute, that it may easily be overlooked.

The figures and descriptions, quoted by LINNÆUS, induce us to consider this species as the *sanguinale*, rather than his specific description, which certainly does not well accord with our plant. The *vaginæ punctatæ* are not mentioned either by HALLER, SCHEUCHZER, or SCOPOLI; the hairs certainly issue from little prominent points, scarcely visible, unless magnified. If LINNÆUS means these, they are, in our apprehension, too minute to form a specific character on.

According to the observations of botanic writers, this species is very universal, being found not only in *Europe*, but *Asia* and *America*; nevertheless, it is not general throughout *England*. It is said to grow about *Elden* in *Suffolk*, at *Witchingham* near *Norwich*, near *Martha's Chapel* by *Guildford*, and in the Gardener's grounds near *Battersea*; in the latter place I found it last *September*, in great plenty; but in one part of the fields only, viz. among the *French* beans, and on the Asparagus beds, in the Western corner of the fields, at no great distance from the Church.

MATTHIOLUS relates, that in *Carniola* the seeds are collected for food, but this is flatly contradicted by SCOPOLI; indeed, for this purpose, it appears much inferior to many others of the same genus.

Its name of *sanguinale* is not taken from its colour, but from an idle trick which the boys in *Germany* have of pricking their nostrils with the spiculæ of this grass, till they draw blood.

PANICUM CRUS GALLI. LOOSE PANIC-GRASS.

PANICUM *Lin. Gen. Pl.* TRIANDRIA DIGYNIA.
    Cal. 3-valvis : valvula tertia minima.
PANICUM *Crus galli* spicis alternis conjugatisque, spiculis subdivisis, glumis aristatis hispidis, rachi
    quinquangulari. *Lin. Syst. Vegetab.* p. 90. *Spec. Pl.* p. 83.
PANICUM spica remota, setis nullis. *Haller Hist.* n. 1544.
PANICUM *Crus galli. Scopoli Pl. Carn.* n. 70.
GRAMEN paniceum spica divisa. *Baub. Pin.* 8.
GRAMEN paniceum, spica divisa, aristis longis armata. var. β. *Bauh. Pin.* 8.
PANICUM sylvestre Herbariorum. *Parkins.* 1154.
PANICUM *vulgare. Ger. emac.* 85. *Raii Syn.* p. 394. Panick Grass with a divided Spike. *Hudson*
    Fl. Angl. ed. 2. p. 24.

RADIX annua, fibrosa.

CAULES plerumque plures ex una radice, primo procumbentes, seu obliqui, demum suberecti, pedales aut bipedales, tribus aut quatuor geniculis distincti.

FOLIA in humilioribus tres quatuorve uncias longa, lineas duas aut tres lata, in procerioribus semipedalia, vel etiam pedalia fere, lineas quatuor, aliquando et sex semptemve lata, acuminata, carinata, laevia, marginibus minutissime denticulatis et ad basin circa orem vaginae pilosis ; *Membrana* nulla ; *Vagina* magna, striata, compressa.

FLORES paniculati.

PANICULA palmaris, et ultra, e pluribus spicis composita, pallide virescens.

SPICÆ saepe ad duodecim, crassae, teretiusculae, plerumque simplices et alternae, etiam ramosae et oppositae, inferioribus sesquiuncialibus, magisque remotis.

RACHIS quinquangularis, angulo quinto obsoleto, ad basin spicarum setosa.

FLORES secundi, turgidi.

CALYX trivalvis, valvula inferiore minore, flosculum recipiente, *fig.* 1. prima et secunda aequalibus, nervosis, mucronatis, hirsutis, altera plana, altera gibbosa. *fig.* 2, 3. *auct.*

COROLLA bivalvis, valvulis aequalibus, glabris, ovatis, altera alteram margine sua amplectente, *fig.* 4. intra valvulam calycis et corollae, membrana tenuis, nunc acuta, nunc emarginata. *fig.* 5, 6.

STAMINA : FILAMENTA tria, capillaria, brevissima ; ANTHERÆ bifurcae, purpurascentes. *fig.* 7.

PISTILLUM : GERMEN subrotundum ; STYLI duo brevissimi ; STIGMATA plumosa, purpurascentia. *fig.* 8.

SEMEN majusculum, nitidum, glumis corollaceis tectum, hinc convexum, inde planum, *fig.* 9, 10. denudatum, *fig.* 11, 12, 13.

ROOT annual and fibrous.

STALKS generally several proceed from the same root, at first procumbent or oblique, finally nearly upright, from one to two feet in height, furnished with three or four joints.

LEAVES in the more humble plants three or four inches long, and from two to three lines in breadth ; in the taller plants six inches, or nearly a foot in length, four lines, and sometimes six or seven broad, tapering to a point, keeled, smooth, the edges very finely toothed, and about the mouth of the sheath hairy ; *Membrane* none ; *Sheath* large, striated and compressed.

FLOWERS in a panicle.

PANICLE a hand's breadth or more in length, of a pale green colour, composed of many spikes.

SPIKES often twelve in number, thick, roundish, generally simple and alternate, sometimes also branched and opposite, the lowermost an inch and a half in length, longer and farther asunder than any of the rest.

RACHIS having five angles, the fifth less perceptible than the others, bristly or hairy at the base of the spikes.

FLOWERS growing one way and turgid.

CALYX composed of three valves, the lowermost or third valve very small, receiving the floscule, *fig.* 1. the first and second equal, rib'd, pointed, hirsute, the one flat, the other gibbous. *fig.* 2, 3. *magnified.*

COROLLA composed of two valves, the valves equal, smooth, ovate, the one with its margin embracing the other, *fig.* 4. betwixt the valve of the calyx and the corolla a thin membrane is observable, which is sometimes notched, and sometimes pointed. *fig.* 5, 6.

STAMINA : three very slender and short FILAMENTS ; ANTHERÆ forked and purplish. *fig.* 7.

PISTILLUM : GERMEN roundish ; STYLES two, very short ; STIGMATA feathered and purplish. *fig.* 8.

SEED rather large, shining, covered by the valves of the corolla, round on one side, and flat on the other, *fig.* 9, 10. taken from its covering. *fig.* 11, 12, 13.

Near the same spot where the *Panicum sanguinale* is found, this rare species also makes its appearance, a space of about twenty yards square was last autumn covered chiefly with this grass, and the *Panicum viride* ; in other parts of the fields it more rarely occurs ; being an annual it may vary its particular place of growth, but by a diligent search will probably always be found in some part of *Battersea Fields* ; Mr. RAY mentions it as having been found between *Deptford* and *Greenwich* in Garden Ground, also in a lane by the *Neat-house Gardens Chelsea*, and by the rivulet side near *Petersfield, Hampshire.* He is surely mistaken in saying *aquosis delectatur.* Mr. HUDSON describes it as growing near *Martha's Chapel, Guildford.*

Botanists make two principal varieties of it, viz. one with awns, and another without ; the latter, which is the state in which we have figured it, most frequently occurs ; it varies also in its colour, being sometimes found with spikes of a purplish hue.

It has a very great affinity to the *Panicum Crus corvi*, a foreign grass, but is too singular to be mistaken for any *English* Panic.

Affording a large quantity of seeds of considerable magnitude, it is sought for with great avidity by sparrows and other small birds.

Panicum Crus Galli

Eriophorum polyſtachion.   Many-Headed Cotton-Graſs.

ERIOPHORUM *Lin. Gen. Pl.* TRIANDRIA MONOGYNIA.

> *Glumæ* paleaceæ, undique imbricatæ. *Cor.* o. *Sem.* 1. *Land* longiſſimâ cinctum.

Raii. Syn. Gen. 23. HERBÆ GRAMINIFOLIÆ NON CULMIFERÆ FLORE IMPERFECTO SEU STAMINEO.

ERIOPHORUM *polyſtachion* culmis teretibus, foliis planis, ſpicis pedunculatis.' *Lin. Syſt. Vegetab.* p. 87. *Sp. Pl.* p. 76. *Fl. Suec.* n. 49.

ERIOPHORUM foliis planis, ſpicis pendulis. *Haller. hiſt.* n. 1331.

LINAGROSTIS *polyſtachia. Scopoli. Fl. Carn.* n. 66.

GRAMEN pratenſe tomentoſum panicula ſparſa. *C. B. Pin.* 4.

GRAMEN tomentarium. *Ger. emac.* 29.

GRAMEN junceum lanatum, vel juncus bombycinus vulgaris. *Park.* 1271. *Scheuncb. Agroſt. ed. Haller.* p. 306. *Vaill. Bot. Paris.* t. 16. f. 1. 2. *Raii Syn.* p. 433. Cotton-graſs.

> *Lightfoot Fl. Scot.* p. 89.
> *Hudſon Fl. Angl. ed.* p. 21.

RADIX perennis, repens, fuſca, ſeu caſtanei coloris, fibrillis plurimis albis, aut rubentibus inſtructa.

ROOT perennial, creeping, of a brown or cheſnut colour, furniſhed with numerous white or reddiſh fibres.

CULMUS ſæpius ſolitarius, dodrantalis, ſeu pedalis, et ultra, erectus, teres, lævis, geniculis duobus parum extantibus plerumque notatus, vaginis foliorum per totam longitudinem tectus.

STALK for the moſt part ſolitary, from nine inches to a foot or more in height, upright, round, ſmooth, for the moſt part furniſhed with two joints which project a little, covered throughout its whole length with the ſheaths of the leaves.

FOLIA *ima* marcida, caſtanea, brevia, lanceolata, ſtriato-reticulata; *ſuperiora* baſi ſua arcté culmum amplectentia, lineas duas ad tres lata, ſpithamæa et ultra, ſenſim attenuata, ſæpe præmorſa, hinc convexa, inde concava, glabra; *ſuprema* planiora, multo breviora, et maniſeſte carinata; *vaginæ* foliorum culmi, æquali ubique fere magnitudine, ubi folium exit paulo laxiores, et fiſturâ membranâ impleta notatæ.

LEAVES *next the root* withered, of a cheſnut colour, ſhort, lanceolate, ſtriated, and marked with ſhort tranſverſe lines, which give them a reticulated appearance, the *ſucceeding leaves*, at their baſe cloſely embracing the ſtalk, from two to three lines in breadth, about ſeven inches or more in length, gradually tapering to the extremity, where they are often bit off, convex on one ſide, concave on the other, and ſmooth, the *uppermoſt leaves* flatter, much ſhorter, and manifeſtly keeled; *ſheaths* of the leaves nearly of an equal thickneſs throughout, where a leaf goes off more looſely connected, and marked with a fiſſure filled by a membrane.

BRACTEÆ tres aut quatuor, longitudine inæquales, baſi vaginantes, culmum terminant, e quarum ſinubus ſpiculæ prodeunt.

FLORAL-LEAVES three or four of unequal lengths, forming ſheaths at bottom terminate the ſtalk, from the alæ of which the ſpiculæ proceed.

SPICULÆ plerumque plores a duabus ad ſeptem, ovatæ, immaturæ erectæ, per ætatem pendulæ.

SPICULÆ for the moſt part ſeveral, from two to ſeven, ovate, firſt upright, afterwards pendulous.

CALYX: *ſpica* undique imbricata: ſquamis ovato oblongis, plano-inflexis, membranaceis, laxis, acuminatis, flores diſtinguentibus. *fig.* 1.

CALYX: a *ſpike* covered on all ſides with imbricated ſquamæ, of an ovate-oblong ſhape, flat and bent in a little, membranous, looſe, running out to a long point, diſtinguiſhing the flowers. *fig.* 1.

COROLLA nulla.

COROLLA wanting.

STAMINA: FILAMENTA tria, capillaria; ANTHERÆ erectæ, oblongæ. *fig.* 2.

STAMINA: three FILAMENTS very fine; ANTHERÆ upright and oblong. *fig.* 2.

PISTILLUM: GERMEN minimum; STYLUS filiformis, longitudine ſquamæ calycis; STIGMATA tria, ſtylo longiora, reflexa. *fig.* 3.

PISTILLUM: GERMEN very ſmall; STYLE threadſhaped, the length of the ſcales of the calyx; STIGMATA three, longer than the ſtyle, turned back. *fig.* 3.

PERICARPIUM nullum.

SEED-VESSEL none.

SEMEN triquetrum, acuminatum, nigrum, villis ſpicâ longioribus inſtructum. *fig.* 4. 5. 6. 7.

SEED three-cornered, pointed, black, furniſhed with hairs which are longer than the ſpike. *fig.* 4. 5. 6. 7.

The Genus *Eriophorum* is in a particular manner diſtinguiſhed from the other genera related to it, by the length of the hairs which envelope the ſeed; and which, when the ſeed is ripe, aſſume the appearance of cotton, whence its name of Cotton-graſs, this cotton is much longer, and produced in greater quantities in the *polyſtachion*, than in the *vaginatum*; and in Germany, and the more northern countries, has been manufactured into various articles of dreſs, paper, and wicks for candles. LINNÆUS, in his *Flora lapponica*, informs us, that in ſome parts of Sweden, the peaſants ſtuff their pillows with it inſtead of feathers, but that in Lapland, where the plant is ſufficiently plentiful, they do not apply it to any ſuch purpoſe, the ſkin of the Rein-deer forming the whole of their bed and its furniture.

In the ſpring, Cattle appear to be very fond of its leaves, as they are generally found cropt, this may ariſe from the ſcarcity or herbage at that ſeaſon of the year, as the plant advances the ſtems are always left untouched; it is in moors and boggy ground only that this plant is found, and in ſuch ſituations it is very plentiful; whole acres being often rendered white as ſnow by it in the months of June and July when in ſeed.

It flowers in April and May, and may be found in Batterſea meadows.

Eriophorum polystachion

Eriophorum *vaginatum*

Eriophorum vaginatum. Single-headed Cotton-Grafs.

ERIOPHORUM *Lin. Gen. Pl.* TRIANDRIA MONOGYNIA. *Glumæ* paleaceæ, undique imbricatæ.
Cor. o. *Sem.* 1. *Land* longiffimâ cinctum.

Raii Syn. Gen. 28. HERBÆ GRAMINIFOLIÆ NON CULMIFERÆ FLORE IMPERFECTO
SEU STAMINEO.

ERIOPHORUM *vaginatum* culmis vaginatis teretibus, fpica fcariofa. *Lin. Syft. Vegetab.* p. 87. *Sp.*
Pl. p. 76. *Fl. Suec. n.* 50.

ERIOPHORUM caule tereti, foliis caulinis vaginalibus, fpica erecta, ovata. *Haller. hift.* n. 1332.

LINAGROSTIS *vaginata. Scopoli Fl. Carn. n.* 66.

JUNCUS alpinus, capitulo lanuginofo, feu fchænolaguros. *Bauh. pin.* 12. *Scheuzch Agroft.* p. 302. t. 7.

JUNCUS alpinus cum cauda leporina, *Bauh. hift.* 2. 514.

GRAMEN juncoides lanatum alterum danicum. *Parkins.* 1271. *Raii Syn.* p. 436. Hares-tail rufh.

Lightfoot Fl. Scot. p. 90.

Hudfon. Fl. Angl. ed. 2. p. 22.

The *Eriophorum vaginatum* is with us a much fcarcer plant than the *polyftachion*, but in fome parts of *Great Britain*, and in other Countries, is equally common.

In its generic characters, excepting the fhortnefs of its *Pappus*, it agrees with the *polyftachion*; it has therefore the fame figures of reference to them; in its fpecific characters, it differs very obvioufly, its root is not creeping but more matted, and its leaves, which are much finer, are confequently more apt to grow in tufts; the vagina of the upper ftem-leaf in particular is remarkably inflated, and it never produces more than one fpike, which is upright.

It is found in fimilar fituations to the other; on the boggy parts of Shirley Common, near Croydon, I have found it in tolerable plenty; it flowers fomewhat earlier than the *polyftachion*, but produces its *pappus* about the fame time.

Sheep are very fond of it, whence in Weftmoreland they call it Mofs-crops. *Raii. Syn.* p. 436.

HOLCUS LANATUS.   MEADOW SOFT-GRASS.

HOLCUS *Lin. Gen. Pl.* POLYGAMIA MONOECIA.

HERMAPHROD. *Cal.* Gluma 1 five 2 flora. *Cor.* Gluma ariſtata. *Stam.* 3.
Styli 2. *Sem.* 1.

MASC. *Cal.* Gluma 2-valvis. *Cor.* o. *Stam.* 3.

Raii Syn. Gen. 27. HERBÆ GRAMINIFOLIÆ FLORE IMPERFECTO CULMIFERÆ.

HOLCUS *lanatus* glumis bifloris villoſis: floſculo hermaphrodito mutico ; maſculo ariſta recurva. *Lin.*
Syſt. Vegetab. p. 760. *Sp. Pl.* 1485. *Fl. Suec.* n. 917.

AVENA diantha, floribus ovatis ; perfecto mutico, imperfecto ariſtato. *Haller. hiſt.* n. 1484.

HOLCUS *lanatus, Scopoli Fl. Carn.* n. 1238.

GRAMEN pratenſe paniculatum molle. *Bauh. pin.* 2.

GRAMEN miliaceum pratenſe molle. *Pet. Conc. Gr.* 224. *Raii Syn.* p. 404. Soft-tufted Meadow-
Graſs.

Hudſon. Fl. Angl. ed. 2. p. 440. *Lightfoot. Fl. Scot.* p. 631.

RADIX perennis, fibroſa, minime repens.

CULMI plures, bipedales, feu tripedales, raro ultra, erecti, quatuor communiter geniculis diſtincti, pubeſcentes, teretes, in quibuſdam locis ad baſin radicantes.

FOLIA pilis mollibus undique veſtita, incana, tres lineas lata, plana, ſuperne ſtriata, inferne carinata ; *vagina* lineis purpureis externe notata, interne nitida ; *membrana* obtuſa, externe piloſa, piliſque ciliata.

PANICULA primo ſpiciformis, mollis, laxus, rubellus, parum nutans, dein erecta, diffuſa, albida.

SPICULÆ biflora, albidæ, villoſulæ, verſus apicem coloratæ.

CALYX : *Gluma* bivalvis, valvula exteriore majore, trinervi, interiore mucronata, minore carinata. *fig.* 1. auct.

FLOS FERTILIS.

COROLLA bivalvis, valvulæ teneræ, virides, nitidæ, muticæ, valvula exteriore majore. *fig.* 2.

STAMINA : FILAMENTA tria, capillaria ; ANTHERÆ oblongæ, bifurcatæ, flavæ. *fig.* 5.

PISTILLUM: GERMEN obovatum ; STYLI duo, ad baſin uſque ramoſi. *fig.* 6.

SEMEN parvum, acuminatum, nitidum glumis corollæ tectum. *fig.* 9. et valvulis calycis incluſum. *fig.* 8.

FLOS STERILIS.

COROLLA bivalvis, valvulæ minores, exteriore ariſtata, ariſta e dorſo valvulæ erumpente, longitudine valvulæ exterioris calycis. *fig.* 3, 4.

STAMINA ut in fertili. *fig.* 5.

PISTILLUM: *Germen* ut in fertili, ſed multo minus ; STYLI duo, ſubulati, ſimplices. *fig.* 7.

SEMEN minimum, abortivum.

ROOT perennial, fibrous, not at all creeping.

STALKS ſeveral, from two to three feet high, ſeldom higher, upright, generally furniſhed with four joints, downy, round, in ſome ſituation taking root at the bottom.

LEAVES covered on every part with ſoft hairs, which give them a greyiſh appearance, about three lines in breadth, flat, above ſtriated, below keeled, the *ſheath* marked externally with purple lines, internally ſhining ; the *membrane* blunt, externally hairy, and edged with hairs.

PANICLE at firſt forms a kind of ſoft, reddiſh ſpike, which droops a little, afterwards becomes upright, ſpreading and whitiſh.

SPICULÆ containing two flowers, whitiſh, ſomewhat villous, coloured towards the top.

CALYX : *a Glume* of two valves, the outer one largeſt, having three ribs, and terminating in a point, the inner one ſmaller and keeled. *fig.* 1. magnif.

FERTILE FLOWER.

COROLLA compoſed of two valves which are tender, green, ſhining, and pointleſs, the outer valve largeſt. *fig.* 2.

STAMINA : three capillary FILAMENTS ; ANTHERÆ oblong, forked at each end, of a yellow colour. *fig.* 5.

PISTILLUM : GERMEN pointed, with two STYLES two, branched quite down to the bottom. *fig.* 6.

SEED ſmall, pointed, and ſhining, covered by the glumes of the corolla, *fig.* 9. and incloſed in the valves of the calyx. *fig.* 8.

BARREN FLOWER.

COROLLA compoſed of two valves, the valves ſmall, the outer one bearded, the awn ariſing from the back of the valve, the length of the outer valves of the calyx. *fig.* 3, 4.

STAMINA as in the fertile flowers. *fig.* 5.

PISTILLUM: the *Germen* as in the fertile flower, but much leſs ; STYLES two, tapering, and ſimple. *fig.* 7.

SEED very minute and abortive.

The *Holcus Lanatus* abounds in moſt meadows, is frequently found by road-ſides, and ſometimes on walls, ſo that it will thrive in almoſt any ſituation. The redneſs of its panicle when juſt opening, joined to the ſoftneſs and hoarineſs of its leaves, render it a very conſpicuous graſs.

HALLER ſpeaks highly of it as food for cattle, calling it *ſpimum pabulum*. We cannot coincide with him in this opinion, nor do the generality of our intelligent farmers and graziers, who condemn it as too ſoft and woolly ; nevertheleſs the ſeed of it (being eaſily collected) is ſometimes ſent up to *London* in great quantities, and ſold for pure graſs-ſeed: but it were better to lay down ground for meadow or paſturage in the uſual way, than fill it with this unprofitable, though pure graſs-ſeed.

Mr. LIGHTFOOT informs us in his *Fl. Scot.* that it is ſometimes uſed to make ropes for the fiſhing-boots.

It is a very diſtinct ſpecies from the *Holcus Mollis* (as we ſhall particularly explain when we treat of that graſs), and flowers in *June* and *July*.

One cannot but lament that LINNÆUS ſhould have ſeparated the *Holcus* from the other graſſes, with which it has ſo great an affinity, and have placed it among the plants of the claſs *Polygamia*, merely becauſe ſome of its flowers were imperfect ; it frequently happening, as HALLER very juſtly obſerves, that the *Triticum, Hordeum*, and ſeveral other graſſes, are in the ſame predicament ; and it was the leſs neceſſary here, as there is an evident piſtillum in the barren flowers, though an imperfect one. *Vid. fig.* 7.

Holcus lanatus.

MILIUM EFFUSUM. MILLET GRASS.

MILIUM *Lin. Gen. Pl.* TRIANDRIA DIGYNIA.

Cal. 2-valvis, uniflorus: valvulis subæqualibus. *Corolla* breviſſima. *Stigmata* penicilliformia.

Raii Syn. Gen. 27. HERBÆ GRAMINIFOLIÆ FLORE IMPERFECTO CULMIFERÆ.

MILIUM *effuſum* floribus paniculatis diſperſis muticis. *Lin. Syſt. Veget.* p. 94. *Sp. Pl.* p. 90. *Fl. Suec.* n. 61.

MILIUM paniculis raris, longiſſime petiolatis. *Haller. Hiſt.* 1525.   o

GRAMEN ſylvaticum, panicula miliacea ſparſa. *Baub. Pin.* 8.

GRAMEN miliaceum. *Lob. icon.* 3. *Ger. emac.* 6. *I. B. II.* 462.

GRAMEN miliaceum vulgare. *Park.* 1153. *Raii Syn.* p. 402. Millet-graſs. *Lightfoot Fl. Scot.* p. 92.

RADIX perennis, repens.

CULMI tenues, tres, quatuorve pedes alti, quatuor communiter geniculis diſtincti, totidemque, vel quinis foliis a geniculis oriundis, veſtiti.

FOLIA palmaria, ſpithamæa, et pedalia, glabra, tenuia, et infirma, ſubtiliſſimè per longitudinem ſtriata, ſuperiuà et infernà parte aſpera, marginibus etiam, ſi deorſum ſtringantur, aſperis donata, tres, quatuorve lineas lata, ſenſim in acutum mucronem terminata. *Vaginæ* ſtriatæ, glabræ, ad internam foliorum baſin in membranulam tenuem, plerumque laciniatam terminatæ.

PANICULA palmaris, frequentius tamen ſpithamæa, pedalis et longior quandoque, ſuberecta, diffuſa, laxa.

RAMI paniculæ, capillares, flexuoſi.

CALYX: *Gluma* uniflora, bivalvis, acuminata, valvulis æqualibus, lævibus, ovatis, acutis. *fig.* 1. 2.

COROLLA bivalvis, calyce minor: valvulæ ovatæ, obtuſiuſculæ, altera minore. *fig.* 3. 4.

STAMINA: FILAMENTA tria, capillaria, corollâ longiora. ANTHERÆ primo oblongæ, demum bifurcæ, flavæ. *fig.* 5.

PISTILLUM: GERMEN ſubrotundum, viride, glabrum; STYLI duo reflexi, plumoſi, albi. *fig.* 6.

SEMEN unicum, tectum, ſubrotundum, nitidum. *fig.* 7.

ROOT perennial, and creeping.

STALKS ſlender, three or four feet high, commonly furniſhed with four joints, and cloathed with as many or five leaves, ariſing from the joints.

LEAVES from four to ſeven inches or a foot in length, ſmooth, thin and weak, very finely ſtriated through their whole length, the upper and under ſide as well as the edges rough if drawn backward through the hand, three or four lines in breadth, terminating gradually in a fine point. *Sheath* ſtriated, ſmooth, at the inner baſe of the leaf terminating in a membrane which is often jagged.

PANICLE four inches in length, but more frequently a ſpan, a foot, or more, nearly upright, ſpreading and looſe.

BRANCHES of the panicle very fine, and crooked.

CALYX: a *Glume* of one flower, and two valves, pointed, the valves equal, ſmooth, ovate, and pointed. *fig.* 1. 2.

COROLLA compoſed of two valves, ſmaller than the calyx: the valves ovate, bluntiſh, one ſmaller than the other. *fig.* 3. 4.

STAMINA: three FILAMENTS, very fine, longer than the corolla. ANTHERÆ firſt oblong, then forked at each end, of a yellow colour. *fig.* 5.

PISTILLUM: GERMEN roundiſh, green, ſmooth. STYLES two, turned back, feathered and white. *fig.* 6.

SEED ſingle, encloſed, roundiſh, and ſhining. *fig.* 7.

The graſs here figured is the only one we have of the genus *Milium*. It is diſtinguiſhed from the *Panics*, to which it has the greateſt natural affinity, by having a calyx of two valves only; the height it uſually attains, the particular ſituation in which it is found, joined to the delicacy of its panicle, eminently diſtinguiſh it from all our other graſſes.

It abounds in many of the woods about town, particularly in *Charlton Wood*, and flowers in *May* with the Lily of the Valley and Hare-bell.

It has a creeping root, and grows readily in a ſhady ſituation.

Milium effusum.

.

Scabiosa arvensis.

SCABIOSA ARVENSIS. FIELD SCABIOUS.

SCABIOSA *Lin. Gen. Pl.* TETRANDRIA MONOGYNIA.

Cal. *communis* polyphyllus; *proprius* duplex, superus. *Recept.* paleaceum sive nudum.

Raii Syn. Gen. 8. HERBÆ FLORE COMPOSITO DISCOIDE, SEMINIBUS PAPPO DESTITUTIS, CORYMBIFERÆ DICTÆ.

SCABIOSA *arvensis* corollulis quadrifidis radiantibus, foliis pinnatifidis incisis, caule hispido. *Lin. Syst. Vegetab.* p. 121. *Sp. Plant.* p. 143. *Fl. Suecic.* n. 117.

SCABIOSA foliis petiolatis, ovato-lanceolatis, dentatis, superioribus semipinnatis. *Haller. Hist.* 206.

SCABIOSA arvensis. *Scopoli Fl. Carn.* n. 135.

SCABIOSA pratensis hirsuta quæ officinarum. *Bauh. pin.* 269.

SCABIOSA major vulgaris. *Ger. emac.* 719.

SCABIOSA vulgaris pratensis. *Parkins.* 484. *Raii Syn.* p. 191. Common Field Scabious. *Hudson. Fl. Angl. ed. II.* p. 62. *Lightfoot Fl. Scot.* p. 114. *Oeder Fl. Dan.* t. 447.

RADIX perennis, ramosa, sublignosa, difficulter evulsa.

ROOT perennial, branched, somewhat woody, with difficulty pulled up.

CAULIS pedalis vel ultra, teres, simplex seu ramosus, scaber, superne nudus, pubescens, inferne nigro punctatus, hispidus: pilis albidis.

STALK a foot or more in height, round, simple or branched, rough, above naked of leaves, and downy below, dotted with black and hispid: the hairs whitish.

FOLIA opposita, hispida, acuta, inferiora integra, ovali-oblonga, remote serrata; superiora sessilia, amplexicaulia, pinnatifida: laciniis linearibus, oppositis, subserratis; intermedia duplo majore, lanceolata, utrinque attenuata, in medio serrata.

LEAVES opposite, hispid, pointed; the lower ones entire, of an oval oblong shape, remotely serrated; the upper ones sessile, embracing the stalk, and pinnatifid; the segments linear, opposite, slightly serrated; the middle segment twice the size of the others, lanceolate, tapering at each extremity, and serrated in the middle.

FLORES terminales, longius pedunculati, solitarii.

FLOWERS terminal, standing on long foot-stalks.

CALYX *communis* polyphyllus, imbricatus, foliolis ovatis, acutis, pubescentibus, ciliatis, plano-patentibus; interioribus paulo minoribus.

CALYX *common to all the florets* composed of many leaves, imbricated, the leaves ovate, pointed, downy, edged with hairs, flat and spreading; the innermost somewhat the smallest.

COROLLA composita hemisphærica, dilute violacea, radiata; *propria radii* tubulata, longitudine calycis, intus villosa, quadrifida: laciniis erectis, oblongis, obtusis, inæqualibus; exteriore paullo majore; duabus lateralibus, oppositis, æqualibus; intima duplo minore, *fig.* 1.; *disci* minor, ore quadrifido, obtuso, inæquali.

COROLLA compound, hemispherical, of a pale violet colour, radiate; the florets in the *circumference* tubular, the length of the calyx, villous within, divided into four segments which are upright, oblong, obtuse and unequal; the outermost somewhat the largest; the two side ones opposite, and equal; the innermost twice as small, *fig.* 1.; the *central florets* smaller, the mouth divided into four, obtuse, unequal segments.

STAMINA: FILAMENTA quatuor, setacea, tubo corollæ inferne adnata, longitudine corollæ; ANTHERÆ exstantes, lineares, incumbentes, corollæ concolores. *fig.* 2.

STAMINA: FILAMENTS four, tapering, growing to the lower part of the tube of the corolla, and of the same length as the corolla. ANTHERÆ projecting, linear, incumbent, of the same colour as the corolla. *fig.* 2.

PISTILLUM: GERMEN inferum, tetragonum, villoso-hispidum, coronatum *pappo* campanulato, villoso-setaceo, cinereo. STYLUS cylindricus, superne incrassatus, corolla longior, erectus. STIGMA exsertum, clavatum, emarginatum. *fig.* 3.

PISTILLUM: GERMEN below the corolla, four-cornered, covered with numerous stiffish hairs and crowned with a bell-shaped *pappus* formed of numerous ash-coloured bristles. STYLE cylindrical, thickened above, longer than the corolla, upright. STIGMA projecting, club-shaped, with a notch. *fig.* 3.

RECEPTACULUM barbatum, pilis germinibus brevioribus.

RECEPTACLE bearded, the hairs shorter than the germina.

SEMEN villosum, subtetragonum, pappo villoso-setaceo coronatum. *fig.* 4.

SEED villous, somewhat four-cornered, crowned with a bristly villous down or pappus. *fig.* 4.

The *Scabiosa arvensis* is a very common plant, both in Corn-fields and Meadows. In the former it is undoubtedly a troublesome weed; in the latter it frequently forms a great part of the pasturage, and being a hardy plant, producing a large quantity of foliage, which is not refused, according to LINNÆUS's experiments by *Kine, Horses,* or *Sheep,* it may perhaps be considered rather as useful.

Dr. RUTTY, in his Materia Medica, remarks, that the leaves have sometimes been described as inodorous and insipid; but, on a more accurate examination, they are found to be bitterish, with some degree of acrimony and astringency. Medicinally this species, as well as the *succisa,* has been recommended internally in Coughs, Asthmas, malignant Fevers, Lues venerea, Epilepsy, &c.; and externally in the Scurvy, Itch, Scabies, Tetters, &c.; and may be used in substance, infusion, decoction, or any manner of way; but, as Dr. LEWIS observes, the present practice has little dependance on it.

It flowers in July and August, varies much in the divisions of its leaves, and is sometimes found with white flowers. The blossoms, and indeed the whole plant is much larger than the *Scabiosa succisa*; its leaves are more jagged. It flowers much earlier, and it affects a drier situation.

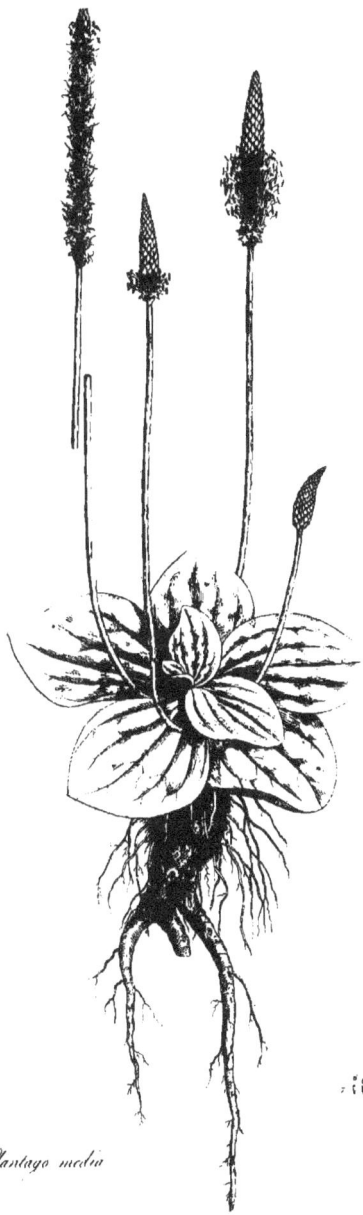

Plantago media

PLANTAGO MEDIA. HOARY PLANTAIN.

PLANTAGO *Lin. Gen. Pl.* TETRANDRIA MONOGYNIA.

Cal. 4-fidus. Cor. 4-fida: limbo reflexo. Stamina longiffima. Capf
2-locularis, circumfciffa.

Raii Syn. Gen. 22. HERBÆ VASCULIFERÆ FLORE TETRAPETALO ANOMALÆ.

PLANTAGO *media* foliis ovato-lanceolatis pubefcentibus, fpica cylindrica, fcapo tereti. *Lin. Syft.
Vegetab.* p. 131. *Sp. Pl.* p. 163. *Fl. Suec.* n. 130.

PLANTAGO foliis fubhirfutis, ellipticis, fpica cylindrica denfa. *Haller. Hift.* n. 659.

PLANTAGO *media. Scopoli Fl. Carniol.* 162.

PLANTAGO latifolia incana. *Bauhin. Pin.* 189.

PLANTAGO major incana. *Parkins.* 493.

PLANTAGO incana. *Ger. emac.* 419. *Raii Syn.* p. 314. Hoary Plantain, or Lamb's Tongue,
Hudfon Fl. Angl. ed. 2. p. 63. *Lightfoot Fl. Scot.* p. 117.

RADIX perennis, craffitie digiti aut pollicis, fubconica, apice in crura aliquot divifa, extus nigricans, plurimis fibrillis inftructa.

ROOT perennial, the thicknefs of one's finger or thumb, fomewhat conic, dividing at the top into a few branches, externally of a blackifh colour, and furnifhed with numerous fibres.

FOLIA ovata, breviffime petiolata, fupra terram expanfa, interioribus fenfim minoribus, quinque-nervia, fubrugofa, utrinque pubefcentia, integerrima.

LEAVES ovate, ftanding on very fhort foot-ftalks, expanded on the ground, the innermoft gradually fmalleft, having five ribs, fomewhat wrinkly, downy on both fides, and entire at the edges.

SCAPI plures, teretes, infra folia prodeuntes, fpithamaei, aut pedales, erecti, pubefcentes, pube fuperne erecta, adpreffa.

FLOWERING-STEMS feveral, round, proceeding from below the leaves, from feven inches to a foot in height, upright, downy, the hairs on the upper part of it upright, and preffed to the ftalk.

SPICÆ florum cylindricæ, pollicares aut palmares.

SPIKES of the flowers cylindrical, from one to four inches in length.

BRACTEA, feu fquamula lanceolata, concava, margine membranacea, fingulo flofculo fubjicitur, longitudine calycis.

FLORAL-LEAF, a floral-leaf or lanceolate fmall hollow fcale, membranous at the edge, and of the length of the calyx, is placed under each flower.

CALYX: PERIANTHIUM quadripartitum, erectum, perfiftens; laciniis ovatis, acutiufculis, membranaceis, nervo viridi infignitis. *fig.* 1.

CALYX: a PERIANTHIUM deeply divided into four fegments, erect and permanent; the fegments ovate, a little pointed, membranous, and marked with a green rib. *fig.* 1.

COROLLA monopetala, perfiftens, tabefcens. *Tubus* cylindraceus, bafi globofus. *Limbus* quadripartitus, depreffus, laciniis ovatis, acutis. *fig.* 2.

COROLLA monopetalous, permanent, withered. *Tube* cylindrical, with a globular bafe. *Limb* divided into four fegments, which are preffed downwards, ovate and pointed. *fig.* 2.

STAMINA: FILAMENTA quatuor, capillaria, erecto-patentia, calyce triplo longiora, purpurafcentia. ANTHERÆ albæ, incumbentes, una extremitate bifidâ, altera mucronatâ. *fig.* 3.

STAMINA: four FILAMENTS very flender, fomewhat fpreading, thrice the length of the calyx, of a purplifh colour. ANTHERÆ white, laying acrofs the filaments, one end bifid, the other pointed. *fig.* 3.

PISTILLUM: GERMEN ovatum. STYLUS filiformis, pilofus, ftaminibus brevior. STIGMA fimplex. *fig.* 4.

PISTILLUM: GERMEN ovate. STYLE thread-fhaped, hairy, fhorter than the ftamina. STIGMA fimple. *fig.* 4.

PERICARPIUM: CAPSULA ovalis, circumfciffa, difperma. *fig.* 6.

SEED-VESSEL: an oval CAPSULE, dividing horizontally in the middle, and containing two feeds. *fig.* 6.

SEMINA bina, hinc convexa, inde plano concava. *fig.* 7.

SEEDS two together, convex on one fide, and plano concave on the other. *fig.* 7.

This fpecies of Plantain has a large root when fully grown, which penetrates deep into the earth, and being fupplied with numerous lateral fibres, it fupports itfelf in the moft fcorching feafons, when the plants around it are frequently burnt up. It is alfo one of thofe plants which are not deftroyed by repeated mowing, as moft lawns and grafs plats fufficiently teftify.

It may be diftinguifhed from the common Plantain by the leaves being fmaller, and hoary, ftanding on fhorter foot-ftalks, lying clofe to the ground, and having no notches on the edges; by its fpikes being fhorter, its filaments longer, its antheræ whiter and more fhowy, and, if any other difference were wanting, we might add, that its capfules, inftead of many, contain only two feeds, as in the *lanceolata.*

About *London* it is not fo common as either the *lanceolata* or *major*; but where the foil is chalky no plant occurs more frequently. It flowers from *June* to *Auguft.*

Sheep, Goats, and Swine, eat it; Kine and Horfes refufe it. *Lin. Pan. Suec.*

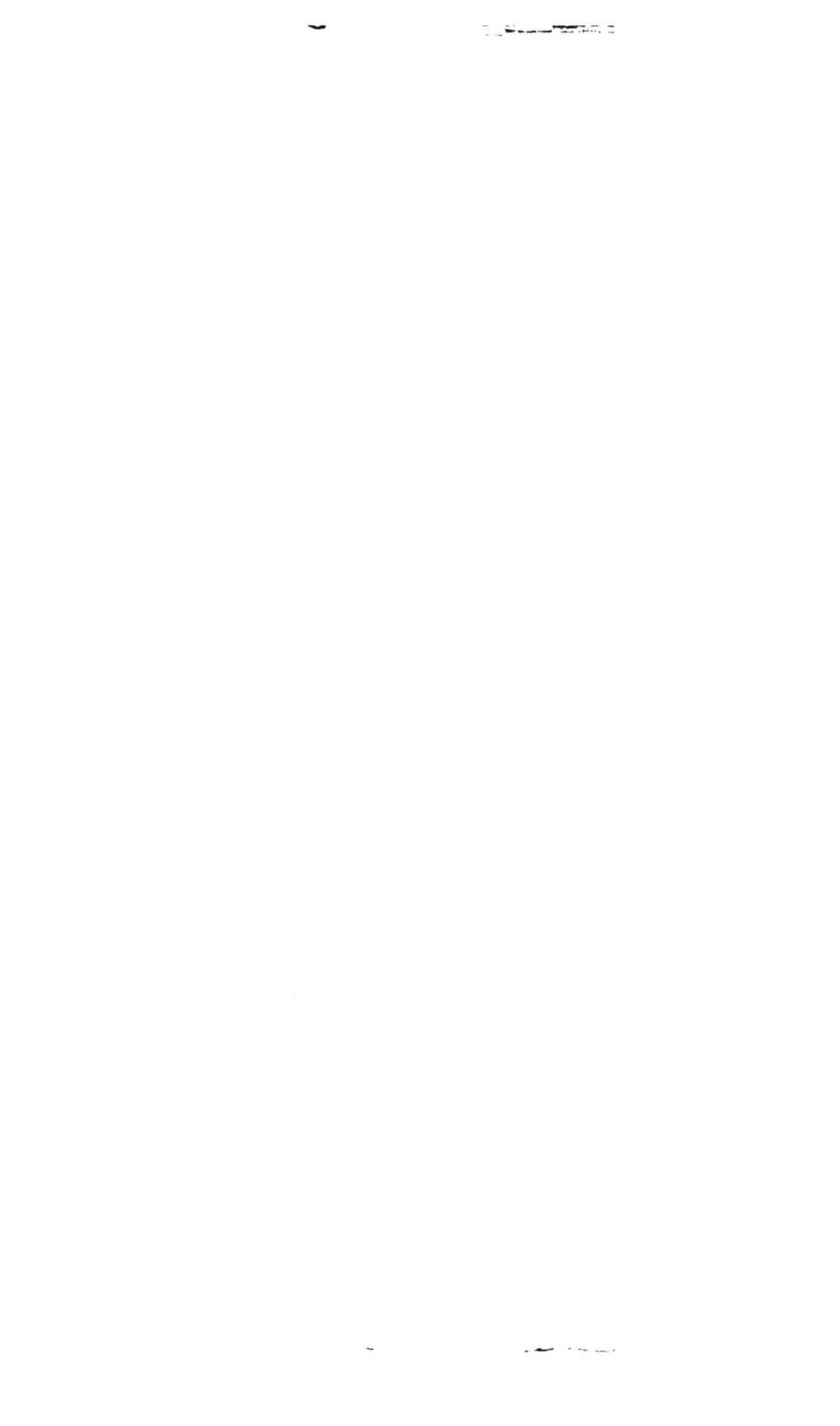

ASPERULA ODORATA. WOODRUFF.

ASPERULA *Lin. Gen. Pl.* TETRANDRIA MONOGYNIA.

  Cor. 1-petala, infundibuliformis. *Semina* 2, globofa.

Raii Syn. Gen. 12. HERBÆ STELLATÆ.

ASPERULA *odorata* foliis octonis lanceolatis, florum fafciculis pedunculatis. *Lin. Syft. Vegetab.* p. 125. *Sp. Pl.* p. 150.

ASPERULA caule erecto, foliis octonis; petiolis ramofis erectis, feminibus hirfutis. *Haller. Hift.* n. 728.

GALIUM odoratum. *Scopoli Fl. Carn.* n. 158.

ASPERULA feu Rubeola montana odora. *Baub. Pin.* 334.

ASPE ULA. *Ger. emac.* 966.

ASPERULA aut Afpergula odorata. *Parkins.* 563. *Raii Syn.* p. 225. Woodroof or Woodruffe. *Hudfon Fl. Angl. ed.* 2. p. 66. *Lightfoot Fl. Scot.* p. 115. *Order Fl. Dan.* t. 562.

RADIX perennis, tenuis, articulatus, flavefcens, paulo infra terram repens, et plurimas fibrillas dimittens.

CAULIS erectus, fpithamæus, plerumque fimplex, tetragonus, quadrifulcatus, glaber.

FOLIA plerumque octona, verticillata, lanceolata, mucronata, glabra, fpinulis ciliata. *fig.* 1.

CORYMBUS terminalis, nudus, erectus.

CALYX nullus.

COROLLA monopetala, alba, infundibuliformis. *Tubus* brevis, germini inlidens. *Limbus* quadripartitus, tubo longior, laciniis lanceolatis, patentibus, craffiufculis. *fig.* 2.

STAMINA: FILAMENTA quatuor, ad apicem tubi, breviffima, alba. ANTHERÆ albidæ, longitudine filamentorum, oblongæ, fublineares, incumbentes. *fig.* 3.

PISTILLUM: GERMEN inferum, viride, fubrotundum, utrinque compreffum, obfolete didymum, hifpidum. STYLUS filiformis, albus, bifidus, antheris brevior. STIGMATA duo, globofa, inæqualia. *fig.* 4, 5, 6.

NECTARIUM: *Glandula* parva bafin ftyli cingens.

PERICARPIUM: *Baccæ* duæ, ficcæ, globofæ, coalitæ, hifpidæ. *fig.* 7.

SEMINA folitaria, fubrotunda, magna. *fig.* 8.

ROOT perennial, flender, jointed, of a yellowifh colour, creeping a little below the furface of the earth, and fending down numerous fmall fibres.

STALK upright, about a fpan in height, for the moft part perfectly fimple, four-cornered, with a groove on each fide, and fmooth.

LEAVES growing generally eight together in a whirl, lanceolate, terminating in a fmall point, fmooth and edged with fmall fpines. *fig.* 1.

CORYMBUS terminal, naked, and upright.

CALYX wanting.

COROLLA monopetalous, white, funnel-fhaped. *Tube* fhort, fitting on the germen. *Limb* divided into four fegments, longer than the tube, fegments lanceolate, fpreading, thickifh. *fig.* 2.

STAMINA: four FILAMENTS at the top of the tube, very fhort and white. ANTHERÆ whitifh, the length of the filaments, oblong, fomewhat linear and incumbent. *fig.* 3.

PISTILLUM: GERMEN placed beneath the corolla, of a green colour, roundifh, flattened on both fides, obfcurely double, and hifpid. STYLE filiform, white, bifid, fhorter than the antheræ. STIGMATA two, globular and unequal. *fig.* 4, 5, 6.

NECTARY: a fmall *Gland* furrounding the bafe of the ftyle.

SEED-VESSEL: two, dry, round, hifpid berries united together. *fig.* 7.

SEEDS fingle, large and roundifh. *fig.* 8.

The flowers of *Woodruff* have an agreeable fmell, and the whole plant, when dried, diffufes an odour like that of the *fweet-fcented Vernal-grafs*. Kept among cloaths, it not only imparts to them an agreeable perfume, but, according to LINNÆUS, preferves them from infects.

RAY informs us, that it gives its flavour to vinous liquors; and that the Germans ufe it much for that purpofe.

As a medicinal plant, it is fuppofed to attenuate vifcid humours, and ftrengthen the tone of the Bowels, whence it is recommended in obftructions of the liver and biliary ducts, and by fome in Epilepfies and Palfies: modern practice has neverthelefs rejected it.

It is common in the woods about *London*, efpecially *Charlton Wood*; and flowers in *May* and *June*.

C Asperula odorata.

Cynoglossum officinale

CYNOGLOSSUM OFFICINALE. HOUNDSTONGUE.

CYNOGLOSSUM *Lin. Gen. Pl.* PENTANDRIA MONOGYNIA.

    Cor. infundibuliformis, fauce claufa fornicibus. *Stamina* deprefsa interiour tantum latere ftylo affixa.

Raii Syn. Gen. 13. HERBÆ ASPERIFOLIÆ.

CYNOGLOSSUM *officinale* ftaminibus corolla brevioribus, foliis lato lanceolatis tomentofis feffilibus. *Lin. Syft. Vegetab.* p. 157. *Sp. Pl.* p. 192. *Fl. Suec.* n. 58.

CYNOGLOSSUM foliis ellipticis, lanceolatis, fericeis, caule foliofo. *Haller. Hift.* n. 587.

CYNOGLOSSUM officinale. *Scopoli Fl. Carniol.* 191.

CYNOGLOSSUM majus vulgare. *Baubin. Pin.* 257. *Ger. emac.* 804. *Parkins.* 511. Great Houndftongue. *Raii Syn. ed.* 3. p. 226. *Hudfon Fl. Angl. ed.* 2. p. 80. *Lightfoot Fl. Scot.* p. 133.

RADIX biennis, craffitie digiti feu pollicis, pedalis et ultra, fufiformis, foris nigricans, intus albida.

ROOT biennial, the thicknefs of the finger or thumb, a foot or more in length, tapering, blackifh on the outfide, and whitifh within.

CAULIS bi feu tripedalis, erectus, fulcato-angulatus, villofus, foliofiffimus, fuperne ramofus; *Rami* plurimi, fuberecti, villofi.

STALK two or three feet in height, upright, grooved or angular, villous, very leafy, branched at top; *Branches* numerous, nearly upright and villous.

FOLIA radicalia magna, pedalia et ultra, petiolata, ovata, acuta, fericea hirfutie incana, venofa, caulina, faltem fuperiora feffilia, conferta, fparfa, erecta, lanceolata, bafi latiora.

LEAVES proceeding from the root large, a foot or more in length, ftanding on footftalks, ovate, pointed, covered with a filky down which gives them a greyifh colour, veiny, thofe of the ftalk at leaft the uppermoft ones feffile, numerous, placed irregularly on the ftalk, upright, lanceolate, and broadeft at the bale.

FLORES primo fordide rubentes, demum cærulefcentes, racemofi, fecundi.

FLOWERS at firft of a dull red colour, afterwards becoming blueifh, growing in racemi or long bunches, and hanging all one way.

PEDUNCULI teretes, alterni, pubefcentes.

RACEMI nearly upright, and generally naked.
FLOWER-STALKS round, alternate, and downy.

CALYX: PERIANTHIUM quinquepartitum, foliolis ovato-lanceolatis, erectis, pubefcentibus, obtufiufculis, interne nitidis. *fig.* 1.

CALYX: PERIANTHIUM deeply divided into five fegments, the leaves ovato-lanceolate, upright, downy, bluntifh, fhining on the infide. *fig.* 1.

COROLLA: monopetala, infundibuliformis; *Tubus* cylindraceus, craffus, pallidus, calyce duplo brevior; *Limbus* concavus, quinquefidus, laciniis rotundatis. *fig.* 2, 3.

COROLLA monopetalous, funnel-fhaped, *Tube* cylindrical, thick, of a pale colour, half the length of the calyx. LIMB concave, divided into five roundifh fegments. *fig.* 2, 3.

NECTARIUM: *Squamæ* quinque, purpureæ, antice concaviufculæ, vertice gibbæ, obtufæ, margini tubi infertæ, limbo duplo breviores, conniventes. *fig.* 4. *Faux* corollæ perforata.

NECTARY: five purple *Scales*, anteriorly hollow, gibbous at top, blunt, inferted into the edge of the tube, half the length of the limb, cloling together. *fig.* 4. *Mouth* of the corolla perforated.

STAMINA: FILAMENTA quinque, breviffima; ANTHERÆ oblongæ, virides, fub fquamis nectareis recondita. *fig.* 5.

STAMINA: five FILAMENTS, very fhort. ANTHERÆ oblong, green, hid under the fcaly nectaries. *fig.* 5.

PISTILLUM: GERMINA quatuor, e luteo-virefcentia, glabra; STYLUS fubulatus longitudine ftaminum, perfiftens; STIGMA obtufum, emarginatum. *fig.* 6.

PISTILLUM: GERMINA four, of a yellowifh green colour, fmooth. STYLE tapering, the length of the ftamina, permanent. STIGMA blunt and nicked. *fig.* 6.

PERICARPIUM: CAPSULA quatuor depreffæ, fubrotundæ, fcabræ, non dehifcentes, apice affixæ. *fig.* 7.

SEED-VESSEL: four flat CAPSULES of a roundifh fhape, fomewhat prickly, not opening, fixed by their points. *fig.* 7.

SEMINA folitaria, fubovata, gibba, acuminata, glabra. *fig.* 8.

SEEDS fingle, fomewhat ovate, gibbous, pointed and fmooth. *fig.* 8.

    The leaves of this plant are in fhape thought to refemble a Dog's-tongue, whence its name.

    It grows wild by road fides, and in uncultivated places; and is particularly common about *Charlton*, and in the road to *Lewifham*.

    It flowers in *June* and *July*, and ripens its feeds in *Auguft* and *September*.

    The whole plant has a difagreeable fmell, much refembling that of mice. Its effects are faid to be narcotic; and an inftance is related in the *Hift. Oxon.* 3. 450*, in which the leaves boiled by miftake for thofe of Comfrey, difordered a whole family, and proved fatal to one. BARON HALLER quotes Dr. BLAIR as defcribing a cafe fomewhat fimilar; but in that inftance the plant ufed was not the *Cynogloffum*, but the *Pulmonaria maritima* +; a plant one would not fufpect of being poifonous.

    Houndftongue has been ufed in medicine both internally and externally; but the prefent practice takes no notice of it in any intention.

    Cattle in general diflike it; but the Goat, who with impunity will eat Deadly Nightfhade and Tobacco, is faid fometimes to crop this naufeous plant; it is the natural food of the caterpillar of the fcarlet tyger-moth (*Phalæna Dominula*) which may be found on it in April and May.

    The *Cynogloffum minus folio virente Ger. emac.* 805, is confidered by LINNÆUS merely as a variety of this plant, and defcribed by RAY as growing in the *London* road between *Kelvedon* and *Wittam* in *Effex*, but more plentifully about *Braxfted* by the way fides. It has alfo been obferved in fome fhady lanes about *Wanftead* by Mr. PITTS; at *Southend*, by *Eltham*, plentifully, Mr. *J. Sherard*; by the road fide, about a mile beyond *Waltham-abbey*, towards *Harlow*, Mr. *Newton*; At *Norbury* in *Surrey*, a mile from *Leatherhead*, plentifully. *Merr. Pin.*

---

* " Mulier quædam Oxonienfis, et maritus cum liberis, et quotquot erant ejus familiæ, folia Cynogloffi (pro tenellis Symphyti Effi guftatis) cocta, libere comederunt; et prandio inox omnes ægrè fe habebant, et non multo poft vomitus moleftos infequebantur, demùm ftupore et fomno correpti funt, nec poft horas fere 40 penitus excitati; una autem mortua eft."

+ *Blair's Mifcellaneous Obfervations,* p. 55.

Menyanthes trifoliata

MENYANTHES TRIFOLIATA. BUCKBEAN.

MENYANTHES *Lin. Gen. Pl.* PENTANDRIA MONOGYNIA.

Corolla hirfuta. *Stigma* 2-fidum. *Capf.* 1-locularis.

Raii Syn. Gen. 18. HERBÆ FRUCTU SICCO SINGULARI, FLORE MONOPETALO.

MENYANTHES *trifoliata* foliis ternatis. *Lin. Syft. Vegetab.* p. 164. *Sp. Pl.* 208. *Fl. Suec.* n. 173. *Fl. Lappon.* p. 50.

MENYANTHES foliis ternatis. *Haller Hift.* n. 633.

MENYANTHES *trifoliata. Scopoli Fl. Carn.* n. 212.

TRIFOLIUM paluftre. *Bauh. Pin.* 327.

TRIFOLIUM paludofum. *Ger. emac.* 1194. *Parkinf.* 1212.

TRIFOLIUM fibrinum *Tabern. et Germanorum. Raii Syn.* p. 285. Marfh-Trefoil, Buckbeans. *Hudfon Fl. Angl. ed.* 2. p. 85. *Lightfoot Fl. Scot.* p. 137. *Oeder Fl. Dan.* 541.

RADIX perennis, repens, longa, geniculata, fibrofa.

ROOT perennial, creeping, long, jointed and fibrous.

CAULIS procumbens, variæ longitudinis pro ratione loci, vaginis tectus.

STALK precumbent, various in its length, according to its fituation, covered by the fheaths of the leaves.

FOLIA petiolata, ternata, ovata, obtufa, utrinque glabra, venofa, margine repanda.

LEAVES ftanding on foot-ftalks, growing three together, ovate, obtufe, fmooth on both fides, veiny, the edge waved or ferpentine.

PETIOLI teretes, ftriati, bafi vaginati.

LEAF-STALKS round, ftriated, forming a fheath at the bottom.

SCAPUS fimplex, nudus, e vaginis foliorum natus, erectus, teres, glaber, foliis longior.

SCAPUS, or flowering ftem, fimple, naked, arifing from the fheaths of the leaves, upright, round, fmooth, longer than the leaves.

THYRSUS terminalis, fubpyramidalis, nudus.

THYRSUS terminal, fomewhat pyramidal, naked.

BRACTEÆ ovatæ, acutiufculæ, concavæ.

FLORAL-LEAVES ovate, fomewhat pointed and hollow.

CALYX: PERIANTHIUM monophyllum, quinque partitum, bafi rugofum, laciniis erectis, oblongis, obtufis, lævibus, margine coloratis. *fig.* 1.

CALYX: a PERIANTHIUM of one leaf, deeply divided into five fegments, at bottom wrinkly, the fegments upright, oblong, obtufe, fmooth, and coloured on the edge. *fig.* 1.

COROLLA monopetala, campanulato-infundibuliformis, extus rofea, intus alba ; *Tubus* craffus, calyce longior, quinque-fulcatus ; *Limbus* quinque-partitus, laciniis ovato-lanceolatis, acutiufculis, reflexo-patentibus, intus barbatis, apicibus nudis, barbâ longitudine calycis, filamentofa, alba. *fig.* 2.

COROLLA monopetalous, betwixt bell and funnel-fhaped, externally of a rofe colour, internally white ; *Tube* thick, longer than the calvx, having five grooves ; *Limb* divided into five fegments, which are narrow and pointed, fpreading and turned back, bearded on the infide, the tips naked, beard the length of the calyx, thready and white. *fig.* 2.

STAMINA: FILAMENTA quinque, fubulata, alba, corollæ tubo adnata ; ANTHERÆ purpurafcentes, fagittatæ, apicibus incurvis ; POLLEN flavum. *fig.* 3.

STAMINA: five FILAMENTS, tapering, white, growing to the tube of the corolla ; ANTHERÆ purplifh, arrow-fhaped, the tips bending in ; POLLEN yellow. *fig.* 3.

PISTILLUM : GERMEN ovatum, viride, nitidum ; STYLUS cylindricus, fuperne paululum incraffatus, ftaminibus duplo longior ; STIGMA bilabiatum, flavum, villofum. *fig.* 4.

PISTILLUM: GERMEN ovate, green, fhining ; STYLE cylindrical, above a little thickened, twice the length of the ftamina ; STIGMA compofed of two lips, green and villous. *fig.* 4.

PARKINSON informs us, that in his time this plant was generally called by the name of *Marfh-trefoil*, and fometimes *Marfh-clauer* ; it is now generally known by the name of *Buckbean*, whether this be a corruption of the compound word *Bog-bean*, or of the low Dutch *Bocx boonen*, hoc eft *Phafeolum hircinum*, or whether it be an original Englifh word we fhall not prefume to determine, it being a point on which the learned themfelves are not agreed ; but there is one point in which all who have feen the Buckbean in perfection will at once agree, viz. that it is one of the moft beautiful plants this country can boaft, nor does it fuffer when compared with the *Kalmia's*, the *Rhododendron's*, and the *Erica's* of foreign climes, which are purchafed at an extravagant price, and kept up with much pains and expence, while this delicate native, which might be procured without any expence, and cultivated without any trouble, bloffoms unfeen, and waftes its beauty in the defart air.

It grows abundantly in moft boggy meadows, it will alfo flourifh in ponds and lakes, and may be found in *Battersea Meadows*, particularly about a hundred yards diftant from the *Red Houfe* towards *Cheifea*, alfo plentifully in the marfhes about the ifland of *St. Helena*, near *Rotherhithe*, and no where in greater plenty than in the marfhes about *Staines*, in many of which it is the principal plant. It flowers in May and June.

To

To such as wish to have this plant flower with them in perfection, I would recommend the following mode of cultivation : collect the roots of the plant either in spring or autumn, put them in a large pot (having a hole at the bottom) filled with bog earth, immerse the pot about two-thirds of its depth in water, in which it should continue ; the advantage of this method is, that when the plant is coming into flower it may be brought into any room and placed in a pan of water, where it will continue to blossom for two or three weeks. A single root which I treated in this manner, planted in the spring, produced the ensuing May eight flowering stems, many of which had fifteen or sixteen blossoms on them.

LINNÆUS, in his invaluable *Flora Lapponica*, has several curious observations on the Buckbean, he informs us that the inhabitants of *some parts of Lapland*, and also of *Westrobothnia*, draw out the roots, which grow plentifully in the rivulets, and for want of other fodder give them to their cattle, who consume them entirely ; also that, in times of severe scarcity, the miserable inhabitants mix the powder of the dried roots with a small quantity of meal and convert them into bread, which, he observes, is very bitter and extremely nauseous ; that it was a general practice with the peasants of *Westrogothia*, in brewing, to substitute the bitter leaves of this plant for the hop, and that they were equally efficacious in preventing the beer from becoming sour ;—he concludes his remarks by observing, that BARTHOLIN, SIMON PAULI, and others, have exclaimed much on the scurvy of the northern regions, arising from cold, and of the profusion of antiscorbutic plants to be met with in those countries, among which they enumerate the Buckbean as a principal one ; LINNÆUS however asserts, that out of the great number of Laplanders he had seen not one was affected with that disease, although they lived in the coldest habitable climate, and used no vegetable for their ordinary food, not even bread. On the contrary, he had observed the scurvy to be one of the most common diseases of those who inhabited the countries adjacent.

Many physicians have a high opinion of its medical virtues.

" It is of subtle penetrating parts, a good *diuretic* and *antiscorbutic*, whence it is also of great use to people
" afflicted with *rheumatic pains*. An *infusion* of the dried leaves may either be drank like tea, or they may be in-
" fused in white wine. It is a very intense *bitter*, and at first not very agreeable. *Dr. Deering Cat. Stirp.*

" Marsh-trefoil is an efficacious aperient and deobstruent, promotes the fluid secretions, and, if liberally taken,
" gently loosens the belly. It has of late gained great reputation in scorbutic and scrophulous disorders ; and its
" good effects in these cases have been warranted by experience ; inveterate cutaneous diseases have been removed
" by an infusion of the leaves drank to the quantity of a pint a day, at proper intervals, and continued some
" weeks." *Lewis's Disp.* p. 242.

It is also extolled for its efficacy in removing a variety of other diseases, as the periodical head-ach, ague, protracted intermittents, jaundice, dropsy, wandering gout, worms, &c.—but we forbear saying more of it on this head, least its real virtues should be called in question.

Dr. TANCRED ROBINSON asserts, that sheep are cured of the rot by being driven to seed it much where this plant abounds ; this, if true, would prove a most valuable discovery.

C

SYMPHYTUM *Lin. Gen. Pl.* PENTANDRIA MONOGYNIA.

Corollæ limbus tubulato ventricofus : fauce claufa radiis fubulatis.

Raii Syn. Gen. 13. HERBÆ ASPERIFOLIÆ.

SYMPHYTUM *officinale* foliis ovato-lanceolatis decurrentibus. *Lin. Syst. Vegetab.* p. 158. *Sp. Pl.* 195. *Fl. Suec.* n. 165.

SYMPHYTUM *Haller. Hist.* n. 600.

SYMPHYTUM *Scopoli Fl. Carn.* n. 195.

SYMPHYTUM Confolida major. *Baub. pin.* 259.

CONSOLIDA major. *Gerard emac.* 806.

SYMPHYTUM majus vulgare. *Parkinfon* 523. *Raii Syn.* p. 230. Comfrey. *Hudfon. Fl. Angl.* ed. 2. p. 81. *Lightfoot Fl. Scot.* p. 134.

RADIX perennis, magna, ramofa, extus nigricans, intus alba, fubdulcenti-infipida, fucco tenaci glutinofo abundans.

ROOT perennial, large, branched, on the outfide blackifh, white within, maukifh, abounding with a flimy juice.

CAULIS bipedalis, erectús, ramofus, teres, fubangulatus, fcaber ; pube rigida, recurva.

STALK about two feet high, upright, branched, round, yet flightly angular, rough; the hairs rigid and bending backwards.

FOLIA alterna, inferne petiolata, fuperne feffilia, decurrentia, ovata, acuta, fpithamæa, etiam pedalia, parum rugofa, venofa, utrinque fcabra, margine fubundulata, ciliata.

LEAVES alternate, the lower ones flanding on footflalks, the upper ones feffile, decurrent, ovate, pointed, feven inches, or even a foot in length, fomewhat wrinkly, veined, rough on both fides, the edges flightly waved, and fringed with hairs.

FLORES ex albo-lutefcentes, raro purpurei, cernui, racemofi, racemis plerumque geminis, involutis, multifloris.

FLOWERS of a yellowifh white colour, rarely purple, drooping, placed on racemi or branches, which ufually grow two together, turn fpirally inwards, and fupport many flowers.

PEDUNCULI racemorum & florum teretes, hirfuti.

PEDUNCLES both of the racemi and flowers, round and very hairy.

CALYX : PERIANTHIUM monophyllum, villofum, profundè quinquefidum : laciniis lanceolatis acutis, carinatis, erectis. *fig.* 1.

CALYX : a PERIANCHIUM of one leaf, hairy, deeply divided into five fegments, which are lanceolate, keel'd and upright. *fig.* 1.

COROLLA infundibuliformis, ex luteo-alba, decidua ; *Tubus* craffus, longitudine calycis, apice extus notatus punctis quinque depreffis, *fig.* 2; *limbus* ovatus, e tubo fenfim ampliato, minutim quinquefidus, laciniis brevibus, rotundatis, revolutis ; *faux* claufa : fquamis nectareis quinque lanceolatis, acutis, margine craffis, ferrulatodentatis, conniventibus, corollâ brevioribus. *fig.* 4, 5.

COROLLA funnel-fhaped, of a yellowifh white colour, deciduous; the *tube* thick, the length of the calyx, marked externally at the top with five fmall depreffions; *fig.* 2. the *limb* ovate from the gradual widening of the tube, divided into five fhort roundifh fegments, which are rolled back; the *mouth* clofed with five long and pointed nectaries, thick at the edge, with numerous teeth-like points, clofing at top, fhorter than the corolla. *fig.* 4, 5.

STAMINA : FILAMENTA quinque, lanceolata, alba, breviufcula; ANTHERÆ oblongæ, apice et bafi bifidæ, lutefcentes, erectæ, fub fquamis nectareis occultatæ. *fig.* 3.

STAMINA : five, lanceolate, white, fhortifh FILAMENTS; ANTHERÆ oblong, bifid both at top and at bottom, of a yellowifh colour, upright, hid by the nectaries. *fig.* 3.

PISTILLUM : GERMEN quadripartitum : lobis fubrotundis, obtufis, viridibus ; STYLUS fubulatus, albus, inter lobos germinis furgens, corollâ paulo longior, apice obliquus ; STIGMA parvum, obtufum. *fig.* 6.

PISTILLUM : GERMEN divided into four lobes, which are roundifh, blunt and green ; STYLE tapering, white, arifing from the middle betwixt the lobes, a little longer than the corolla, with a fmall obliquity at top ; STIGMA fmall and blunt. *fig.* 6.

SEMINA quatuor in fundo calycis, majufcula, angulata, nigricantia, nitida. *fig.* 7.

SEEDS four, in the bottom of the calyx, largifh, angular, blackifh and fhining. *fig.* 7.

The *Comfrey* is a very common plant by river fides, on the edges of wet ditches, and in other moift fituations ; it flowers from June to September.

Its bloffoms are for the moft part of a yellowifh white colour, but in fome parts of *England*, and abroad, they are more commonly purple.

As a medicinal plant the *Comfrey* has been held in high eftimation, its confolidating virtues have however been carried to a ridiculous excefs; the roots, which are full of a glutinous juice, agree in quality with the roots of Marfh-mallow, and hence are recommended, internally, in fpittings of blood, purgings, fluxes, and ulcers of the bladder; externally, by way of poultice to frefh wounds, fractured bones, bad ulcers, bruifes, gouty fwellings, &c.

A decoction of the powdered root, prepared in a particular manner, yields a fine fcarlet colour. HELLOT *teinture*, p. 357.

It is generally left untouched by cattle. I know of no plant, that on being repeatedly cut down, produces fuch a quantity of herbage.

Symphytum officinale.

VINCA MAJOR. GREAT PERIWINKLE.

VINCA *Lin. Gen. Pl.* PENTANDRIA MONOGYNIA.

Contorta. *Folliculi* 2, erecti, *Semina* nuda.

Raii. Syn. Gen. 17. HERBÆ MULTISILIQUÆ SEU CORNICULATÆ.

VINCA major caulibus erectis, foliis ovatis, floribus pedunculatis. *Lin. Syst. Vegetab.* p. 304.

PERVINCA caulibus erectis, foliis ovato-lanceolatis ciliatis, petiolis unifloris. *Haller. hist.* n. 573.

PERVINCA major. *Scopoli Fl. Carn.* n. 174.

CLEMATIS daphnoides major. *Bauh. pin.* 301.

CLEMATIS daphnoides a. Pervinca major. *Ger. emac.* 894.

CLEMATIS daphnoides latifolia, f. Pervinca major. *Parkins.* 380. *Raii Syn.* p. 268. The greater Periwinkle. *Hudson Fl. Angl. ed.* p. 91.

RADIX perennis, fibrosa, fibris albidis seu fuscis.

CAULES florigeri erecti, bipedales et ultra, teretes, lateribus alterne subcompressis, glabri, rubro maculati, cauliculi etiam sunt steriles qui humi repent aut plantas vicinas scandent.

FOLIA opposita, petiolata, ovata, glaberrima, minutim ciliata, remota, sempervirentia.

PEDUNCULI foliis longiores, erecti, teretes, glabri, filiformes, uniflori.

FLORES ampli, pallide cærulei.

CALYX: PERIANTHIUM quinquepartitum, laciniis sublinearibus, ciliatis, longitudine fere tubi corollæ. *fig.* 1.

COROLLA monopetala, hypocrateriformis, tubus inferne cylindraceus, superne latior, lineis quinque insculptus, ore pentagono, albido, limbus horizontalis, quinquepartitus, laciniis extrorsum latioribus, oblique truncatis. *fig.* 2.

STAMINA: FILAMENTA quinque brevissima, inflexa, retroflexa; ANTHERÆ biloculares, introrsum dehiscentes, membrana pilosa incurva terminatæ. *fig.* 3.

PISTILLUM. GERMINA duo, compressa duabus nitidis liquorem melleum copiose effundentibus; STYLUS utrique unus communis, ad basin sensim gracilescens; STIGMATA duo, inferius orbiculatum, planum, superius albissimum, pilosum, membranis antherarum obtecta. *fig.* 4.

ROOT perennial and fibrous, the fibrous whitish or of a brown colour.

STALKS producing the flowers upright, two feet high and upwards, round, the sides alternately somewhat flatten d, smooth, dotted with red, there also other stalks producing no flowers which creep on the ground or climb the neighbouring plants.

LEAVES opposite, standing on footstalks, ovate, smooth, shining, finely edged with hairs, remote from each other, and evergreen.

FLOWER-STALKS longer than the leaves, upright, round, smooth, filiform, each supporting one flower.

FLOWERS large, of a pale blue colour.

CALYX: a PERIANTHIUM divided into five segments, the segments somewhat linear, ciliated, almost the length of the tube of the corolla. *fig.* 1.

COROLLA monopetalous, salver-shaped, the tube below, cylindrical, above broader, having five grooves, the mouth whitish, five cornered, the limb horizontal, divided into five segments which are externally broadest, and obliquely cut off. *fig.* 2.

STAMINA: five FILAMENTS very short, first bent in, and afterwards bent back; ANTHERÆ bilocular, opening inwardly, terminated by a hairy membrane bent in at top. *fig.* 3.

PISTILLUM: GERMINA two, pressed by two shining glands which pour forth plentifully a sweet liquor; STYLE one common to both, gradually tapering to the base; STIGMATA two, the lowermost round and flat, the uppermost very hairy, covered by the membrane of the antheræ. *fig.* 4.

In the *Systema Vegetabilium* of LINNÆUS, the last of his works published under his direction, this plant is suspected to be a variety of the *Vinca minor*, a suspicion for which there appears to be no grounds, and which is contrary to the united opinion of Botanists both antient and modern; the *minor* it is true has many varieties, but they relate to the colour of the blossoms, and leaves, and the multiplication of the former merely, no alteration is produced in the general habit of the plant, not even by long continued culture; HALLER, in his specific character of the *major*, observes, that the leaves are finely edged with hairs, so far as our observation extends this is constant, and may serve, if any difficulty of distinguishing them should arise, to settle it.

The *major*, like the *minor*, is common enough with us in gardens, but rarely met with wild, yet I have noticed it in several places, particularly under Lord Stormont's Park pales, on the left hand side of the road, betwixt Wandsworth and Putney-Common, and in a field near Beckenham, in Kent, where it was certainly in a wild state.

It flowers in May and June.

It is regarded only as an ornamental plant, but should be introduced into the garden with caution, as it encreases very much, and is apt to over-run and injure others.

vinca major.

SAMOLUS VALERANDI. ROUND-LEAVED WATER-PIMPERNEL.

SAMOLUS *Lin. Gen. Pl.* PENTANDRIA MONOGYNIA.

Cor. hypocrateriformis. Stamina munita squamulis corollæ. Capf. unilocularis.

Raii Syn. Gen. 18. HERBÆ FRUCTU SICCO SINGULARI FLORE MONOPETALO.

SAMOLUS *Valerandi. Lin. Syst. Vegetab.* p. 177. *Sp. Pl.* p. 243. *Fl. Suec.* n. 192.

SAMOLUS. *Haller Hist.* n. 707.

ANAGALLIS aquatica, rotundo folio non crenato. *Baubin. Pin.* 252.

SAMOLUS Valerandi. *Baub. Hist.* 3. p. 791.

ALSINE aquatica, foliis rotundis becabungæ. *Morif. Hist.* 2. p. 323. f. 3. t. 24. f. 28.

ANAGALLIS aquatica rotundifolia. *Ger. emac.* 620.

ANAGALLIS aquatica tertia Lobeli folio fubrotundo non crenato. *Parkins*, p. 1237. *Raii Syn.* 283. Round-leaved Water-Pimpernel. *Hudfon. Fl. Angl.* ed. 2. p. 94. *Lightfoot Fl. Scot.* p. 142. *Order Fl. Dan. icon.* 198.

RADIX perennis, fibrofa, alba.

CAULIS dodrantalis, aut pedalis, erectus, rigidulus, teres, glaber, plerumque ramofus.

FOLIA alterna, petiolata, ovata, obtufa, integerrima, nitida, venis paucis, remotis, notata.

PETIOLI breves, lati.

FLORES parvi, albi, fpicati.

PEDUNCULI plurimi, fparfi, fuberecti, uuiflori, teretes, bracturâ lanceolatâ medio inftructi.

CALYX: PERIANTHIUM quinque-partitum, fuperum, bafi obtufum, laciniis erectis, perfiftentibus. *fig.* 1.

COROLLA monopetala, hypocrateriformis. *Tubus* breviffimus, longitudine calycis, patulus. *Limbus* planus, quinque-partitus, obtufus. *Squamulæ* quinque, breviffimæ, ad bafin finus limbi, conniventes, *fig.* 2, 3.

STAMINA: FILAMENTA quinque, brevia, infra laciniam corollæ fingula. ANTHERÆ conniventes, luteæ, intra tubum corollæ. *fig.* 4.

PISTILLUM: GERMEN inferum. STYLUS filiformis, longitudine fere ftaminum. STIGMA capitatum. *fig.* 5.

PERICARPIUM: *Capfula* fubrotunda, calyce cincta, unilocularis, ore quinquedentato, dentibus reflexis. *fig.* 6.

SEMINA plurima, exigua, nigra. *Receptaculum* globofum, magnum. *fig.* 7, 8.

ROOT perennial, fibrous, and white.

STALK a fpan or a foot in height, upright, fomewhat rigid, round, fmooth, and generally branched.

LEAVES alternate, ftanding on foot-ftalks, ovate, obtufe, perfectly entire, fhining, marked with few veins, and thofe diftant.

LEAF-STALKS fhort, and broad.

FLOWERS fmall, white, growing in fpikes.

FLOWER-STALKS numerous, placed in no regular order, nearly upright, each fupporting one flower, round, having a fmall pointed floral-leaf growing from the middle of each.

CALYX: a PERIANTHIUM deeply divided into five fegments, placed above the germen, blunt at the bafe, the fegments upright and permanent. *fig.* 1.

COROLLA monopetalous, falver-fhaped. *Tube* very fhort, the length of the calyx, open. *Limb* flat, deeply divided into five fegments, which are obtufe; five very fhort *Scales* which clofe inward, are fituated at the mouth of this tube. *fig.* 2, 3.

STAMINA: five fhort FILAMENTS placed beneath each fegment of the corolla. ANTHERÆ clofing together, of a yellow colour, within the tube of the corolla. *fig.* 4.

PISTILLUM: GERMEN beneath the calyx. STYLE thread-fhaped, nearly the length of the ftamina. STIGMA forming a little head. *fig.* 5.

SEED-VESSEL: a roundifh *Capfule*, covered by the calyx, of one cavity, the mouth having five teeth, which turn back. *fig.* 6.

SEEDS numerous, fmall and black. *Receptacle* round and large. *fig.* 7, 8.

Of this genus there is at prefent only one known fpecies, and that an inhabitant of every quarter of the globe: neverthelefs, it is found but fparingly with us; but may with certainty be met with on the edges of the wet ditches about *Woolwich, Charlton,* and *Greenwich,* more particularly in the road leading from *New-crofs* Turnpike to the *Ifland of St. Helena, Rotherhithe.*

It flowers in *July.*

No particular ufes are attributed to it.

Samolus *Valerandi.*

Campanula rotundifolia. Heath Bell-flower.

CAMPANULA *Lin. Gen. Pl.* Pentandria Monogynia.

> *Cor.* campanulata, fundo clauſo valvis ſtaminiferis. *Stigma* trifidum.
> *Caps.* infera, poris lateralibus dehiſcens.

Raii Syn. Gen. 18. Herbæ fructu sicco singulari flore monopetalo.

CAMPANULA *rotundifolia* foliis radicalibus reniformibus, caulinis linearibus. *Lin. Syſt. Vegetab.* p. 173. *Sp. Plant.* p. 232. *Fl. Suec.* n. 184.

CAMPANULA foliis ſerratis, radicalibus cordatis, caulinis lanceolatis. *Haller. Hiſt.* 701.

CAMPANULA *rotundifolia. Scopoli Fl. Carn.* n. 124.

CAMPANULA *minor* rotundifolia vulgaris. *Bauhin pin.* 93.

CAMPANULA rotundifolia. *Ger. emac.* 452.

CAMPANULA minor ſylveſtris rotundifolia. *Parkinſon* 651. *Raii Syn.* p. 277. The leſſer round-leaved Bell-flower. *Hudſon. Fl. Angl. ed.* 2. p. 95. *Lightfoot Fl. Scot.* p. 141.

RADIX alba, craſſiuſcula, modice fibroſa, ſubdulcis, repens.

ROOT white, thickiſh, moderately fibrous, ſweetiſh, and creeping.

CAULES ex una radice plures, pedales et ultra, ſub-erecti, debiles, flexuoſi, teretes, glabri, ſolidi, lacteſcentes, ramoſi.

STALKS; from the ſame root ſeveral, a foot or more in height, ſomewhat upright, but weak and crooked, round, ſmooth, ſolid, milky, and branched.

FOLIA radicalia cordato-reniformia, petiolata, dentato-ſerrata, caulina prope baſin lanceolata, dentata, ſumma linearia, integerrima.

LEAVES of the root ſomewhat heart or kidney ſhaped, ſtanding on footſtalks, toothed, or ſawed, thoſe of the ſtalk near the baſe lanceolate and toothed, near the ſummit linear and entire.

RAMI floriferi, patuli, ſimplices ſeu ramoſi, ſubnudi.

FLOWER-BRANCHES ſpreading, ſimple or branched, almoſt naked.

FLORES perfecte campanulati, cærulei, parum nutantes.

FLOWERS perfectly bell-ſhaped, of a blue colour, and drooping a little.

CALYX: Perianthium quinquepartitum, erectum, glabrum, ſulcatum, perſiſtens, laciniis linearibus. *fig.* 1.

CALYX: a Perianthium divided into five ſegments, upright, ſmooth, grooved, permanent, the ſegments linear. *fig.* 1.

COROLLA monopetala, campanulata, limbo quinque-fido, laciniis acutis, patentibus. *fig.* 2.

COROLLA monopetalous, bell-ſhaped, divided into five ſegments at the brim, which are pointed and ſpreading. *fig.* 2.

STAMINA: Filamenta quinque, capillaria, breviſſima, inſerta valvularum nectarii apicibus; Antheræ filamentis longiores, compreſſæ, primo purpuraſcentes, dein fuſcæ. *fig.* 3, 4, 5.

STAMINA: five very fine, ſhort Filaments inſerted into the tips of the valves of the nectary; Antheræ longer than the filaments, flatten'd, at firſt purpliſh, afterwards brown. *fig.* 3, 4, 5.

PISTILLUM: Germen inferum, ſulcatum; Stylus filiformis; Stigma tripartitum, oblongum, craſſiuſculum, externe villoſum, laciniis revolutis. *fig.* 6, 7, 8.

PISTILLUM: Germen beneath the calyx, grooved; Style thread-ſhaped; Stigma oblong, thickiſh, externally villous, divided into three ſegments which are rolled back. *fig.* 6, 7, 8.

NECTARIUM in fundo corollæ, conſtructum valvulis quinque, acutis, conniventibus, receptaculum tegentibus. *fig.* 5.

NECTARY in the bottom of the corolla, formed of five pointed valves cloſing and covering the receptacle. *fig.* 5.

When the *Campanula rotundifolia* grows among herbage, its radical leaves, which are of a roundiſh figure, at leaſt compared with moſt of the plants of the ſame genus, are ſeldom obſerved, unleſs particularly ſought for, while thoſe of the ſtalk are ſeen by every one to be linear; hence the name of *rotundifolia* to moſt beginners appears abſurd. Linnæus in giving it this name has followed the antient Botaniſts, as will appear from conſulting the ſynonyms.

This plant, as well as the *Epilobium anguſtifolium*, points out to the ſtudent the neceſſity of attending to the following botanic axiom, *thoſe bloſſoms which are on the point of expanſion ſhew the ſtructure of the ſtamina to the moſt advantage, as thoſe which are overblown do that of the ſtigma.*

Subject to the ſame variation in point of ſize with all other plants, it may be found from * two inches to a yard in height, its radical leaves in certain ſituations are found without any notches, in which caſe it is more truly *rotundifolia*; its bloſſoms alſo vary in their colour, being ſometimes found white and ſometimes purple.

The ſtalks and branches, when broken, give out a milky juice, which has a diſagreeable ſmell.

It grows plentifully on heaths, and by the road ſides in barren hilly ſituations, and flowers from June to September.

Having a perennial and a creeping root it is eaſily cultivated in the Garden.

Linnæus ſays a green pigment is prepared from the flowers, but does not inform us in what manner.

* Mr. *Lightfoot* found it of that height in *Scotland.*

Campanula rotundifolia.

CHIRONIA CENTAURIUM. CENTAURY.

CHIRONIA *Lin. Gen. Pl.* PENTANDRIA MONOGYNIA.

Cor. rotata. *Piftillum* declinatum. *Stamina* tubo corollæ infidentia. *Antheræ* demum fpirales. *Pericarp.* 2-loculare

Raii Syn. Gen. 18. HERBÆ FRUCTU SICCO SINGULARI FLORE MONOPETALO.

CHIRONIA *Centaurium.*

GENTIANA *Centaurium* corollis quinquefidis infundibuliformibus, caule dichotomo, piftillo fimplici. *Lin. Syft. Vegetab.* p. 122. *Sp. Plant.* p. 332. *Fl. Suec.* n. 232.

GENTIANA caule dichotomo; floribus infundibuliformibus, ftriatis, quinquefidis. *Haller. Hift.* n. 648.

GENTIANA *Centaurium. Scopoli Fl. Carn.* n. 293.

CENTAURIUM minus *Bauh. Pin.* 278.

CENTAURIUM minus vulgare. *Parkins.* 272.

CENTAURIUM parvum. *Gerard. emac.* 547. *Raii Syn.* p. 286. Small Purple Centory. *Hudfon Fl. Angl.* ed. 2. p. 102. *Lightfoot Fl. Scot.* p. 152.

RADIX annua, fibrofa, lignofa, flavefcens.

ROOT annual, fibrous, woody, and of a yellowifh colour.

CAULIS fpithamæus, et ultra, erectus, plerumque, fimplex, glaber, angulofus.

STALK about feven inches high or more, upright, generally fimple, fmooth, and angular.

FOLIA oppofita, feffilia, glabra, radicalia oblonga, apice obtufa, bafi anguftata, caulina ovato-lanceolata, erecta, trinervia, fuperioribus fæpe incurvis.

LEAVES oppofite, feffile, fmooth, thofe of the root oblong, blunt at the point and narrowed at the bafe; thofe of the ftalk narrow, pointed, upright, three-ribbed, the uppermoft often bent inward.

FLORES rofei, corymbofi, erecti, feffiles.

FLOWERS rofe-coloured, growing in a corymbus, upright, and feffile.

CALYX: PERIANTHIUM monophyllum, quinquefidum, erectum, corollæ fubagglutinatum, perfiftens, laciniis fubulatis, fubtriangularibus, membranâ connexis. *fig.* 1, 2. *auct.*

CALYX: a PERIANTHIUM of one leaf, divided into five fegments, upright, flightly glued to the corolla, permanent, the fegments tapering to a point, fomewhat triangular, connected by a membrane. *fig.* 1, 2. *magnified.*

COROLLA monopetala, infundibuliformis, *tubus* cylindraceus, ftriatus, tenuiffimus, calyce duplo longior, *limbus* quinquepartitus, rofeus, laciniis ovatis, patentibus. *fig.* 3.

COROLLA monopetalous, funnel-fhaped, the *tube* cylindrical, ftriated, extremely thin, twice the length of the calyx, *limb* divided into five fegments, of a rofe-colour, the fegments ovate and fpreading. *fig.* 3.

STAMINA: FILAMENTA quinque, alba, filiformia, ex apice tubi enata. ANTHERÆ oblongæ, incumbentes, flavæ, demum fpiraliter contortæ. *fig.* 4, 5, 6.

STAMINA: five FILAMENTS, white, thread-fhaped, fpringing from the top of the tube. ANTHERÆ oblong, incumbent, of a yellow colour, finally twifted. *fig.* 4, 5, 6.

PISTILLUM: GERMEN oblongum, tubum corollæ implens. STYLUS albus, filiformis, germine dimidio brevior, declinatus, aliquando bifidus. STIGMA craffum, bilobum, villofum, *fig.* 7, 8, 9.

PISTILLUM: GERMEN oblong, filling the tube of the corolla. STYLE white, of equal thicknefs throughout, half the length of the germen, inclining to one fide, fometimes bifid. STIGMA thick, compofed of two lips and villous. *fig.* 7, 8, 9.

PERICARPIUM: CAPSULA oblonga, acuta, nitida, tubo corollæ obtecta, bilocularis, bipartibilis.

SEED-VESSEL: a CAPSULE, of an oblong fhape, pointed, fhining, covered by the tube of the corolla, divifible into two parts, with a cavity in each.

SEMINA numerofa, parva, fubrotunda, flavefcentia.

SEEDS numerous, fmall, roundifh, of a yellowifh colour.

Thofe who have been accuftomed to confider this well-known plant as a *Gentiana*, will be ftartled at feeing it here firft announced as a *Chironia*; but when they come attentively to examine its parts of fructification, they will wonder how they could fo readily acquiefce in joining it to a genus with which its ftructure is wholly irreconcileable. It agrees perfectly with LINNÆUS's character of the genus *Chironia*, the effence of which confifts in its *twifted Antheræ*; and it is worthy of obfervation, that the bloffoms of two of the *Chironia's*, not unfrequently met with in the gardens of the curious, are of the fame colour as the Centaury. Thefe facts have induced me to add a new genus to the Englifh Catalogue, whereby this plant fortunately affumes its proper name *.

The Centaury grows wild in dry and barren fields, on heaths by the fides of hedges, and fometimes in woods, where it ufually acquires a greater height. In the neighbourhood of *Charlton* and *Coombe Woods* it is not unfrequent, and flowers in *July* and *Auguft*.

A variety, with white flowers, is not uncommon.

This herb is extremely bitter, with a difagreeable tafte, whence, Baron HALLER obferves, the ancients called it *fel terræ*, or *gall of the earth*. From an idea, however, that all bitters are good ftomachic medicines, it has acquired no fmall degree of medicinal fame, and is particularly recommended in all weakneffes of the Stomach; alfo in the Jaundice, Green-ficknefs, Worms, Agues, Gout, Scurvy, &c.

It may be given in fubftance to a drachm; in infufion or decoction to two ounces; the extract to a fcruple. Authors have remarked, that it is a plant very difficult of cultivation.

* Centaury has its name ενταυρε or ενταυρον from *Chiron* the Centaur, "Centaurea curatus dicitur Chiron, cum Herculis excepti hofpitio perfractandi arma fagitta excidiffet in pedem: quare aliqui Chironion vocant." *Plin.* l. 25. c. 6. p. 635.

Chironia Centaurium.

Chenopodium hybridum.

CHENOPODIUM HYBRIDUM. THORN-APPLE-LEAVED GOOSEFOOT.

CHENOPODIUM *Lin. Gen. Pl.* PENTANDRIA DIGYNIA.

Cal. 5-phyllus, 5-gonus. *Cor.* o. *Sem.* 1 lenticulare, fuperum.

Raii Syn. Gen. 5. HERBÆ FLORE IMPERFECTO SEU STAMINEO (VEL APETALO POTIUS).

CHENOPODIUM *Hybridum* foliis cordatis angulato-acuminatis, racemis ramofis nudis. *Lin. Syft. Veget.* p. 216. *Sp. Pl.* p. 319. *Fl. Suec.* n. 220.

CHENOPODIUM foliis glabris feptangulis, floribus paniculatis. *Haller. Hift.* n. 1588.

ATRIPLEX fylveftris latifolia, acutiore folio. *Bauh. Pin.* 119.

CHENOPODIUM Stramonii folio. *Vaillant, Paris* 36. *t. 7. f. 2.*

CHENOPODIO affinis, folio lato, laciniato, in longiffimum mucronem procurrente, florum ramulis fparfis. *Raii Hift. Ill.* 123.

BLITUM Aceris folio. *Pet. H. Brit.* 8. 7.

ATRIPLEX odore et folio Stramonii minori tamen. *Lœl. Triumf.* apud fratrem. *Raii Syn.* p. 154. *Hudfon. Fl. Angl.* ed. 2. p. 105.

RADIX annua, fimplex, fibrofa, fibris plurimis, capillaceis, patentibus.

ROOT annual, fimple and fibrous, fibres numerous, capillary, and fpreading.

CAULIS pedalis, ad bipedalem erectus, ramofus, angulatus, glaber.

STALK from one to two feet high, upright, branched, angular, and perfectly fmooth.

FOLIA petiolata, alterna, glabra, fine farinâ, venofa, fubtriangularia, patentia, utrinque plerumque tridentata, acuminata, dentibus magnis, remotis.

LEAVES ftanding on foot-ftalks, alternate, fmooth, without any meal, veiny, fomewhat triangular, fpreading, furnifhed with three teeth on each fide, and running out to a long point, the teeth large, and diftant from each other.

PETIOLI foliis breviores, fubrugofi, inferne convexi, fuperne canaliculati.

LEAF-STALKS fhorter than the leaves, fomewhat wrinkled, convex on the under, and hollow on the upper fide.

FLORES paniculati.

FLOWERS growing in a panicle.

PANICULA ampla, ramofiffima nuda.

PANICLE large, very much branched and naked.

CALYX: PERIANTHIUM pentaphyllum, perfiftens, foliolis ovatis, obtufiufculis, pulverulentis, margine membranaceis.

CALYX: PERIANTHIUM of five leaves and permanent, leaves ovate, fomewhat obtufe, mealy, membranous at the edge.

COROLLA nulla.

COROLLA wanting.

STAMINA: FILAMENTA quinque fubulata, flavefcentia, longitudine calycis. ANTHERÆ didymæ, fubrotundæ, flavæ, *fig.* 2.

STAMINA: FILAMENTS five, tapering, yellowifh, the length of the calyx. ANTHERÆ double, roundifh, and yellow, *fig.* 2.

PISTILLUM: GERMEN orbiculatum. STYLUS bipartitus, brevis. STIGMATA minima, fubreflexa. *fig.* 3.

PISTILLUM: GERMEN round, but fomewhat flattened. STYLE fhort, bipartite. STIGMATA very fmall, turned fomewhat back. *fig.* 3.

SEMEN e nigro caftaneum orbiculare, depreffum, intra calycem.

SEED of a dark chefnut colour, orbicular, flattened, contained within the calyx.

LINNÆUS, when he beftowed on this *Chenopodium* the name *Hybridum*, had an idea, as may be learned from his *Flora Suecica* *, that it was a fpurious plant produced from the *viride*: repeated obfervations would probably have taught him, that this opinion was too haftily adopted, as the *hybridum* has certainly as great pretenfions to be confidered as an original fpecies, as the *viride*, the *album*, or any other. Indeed it is one of thofe *Chenopodiums* which varies the leaft of any; and, befides the form of its leaves, which refemble thofe of the Thorn-apple, and its peculiarly branched and naked panicle of flowers, it has a ftrong and difagreeable fmell, which fome have compared to that of the Thorn-apple: Alfo fown in the garden it produces invariably the likenefs of the original plant.

Of all the *Englifh* plants of this genus, and we have them all (the maritime excepted) growing wild about **London**, the environs of which are in many places peculiarly favourable to their growth, this is by far the fcarceft. I have hitherto difcovered it in one place only, and that fparingly, *viz.* in *Batterfea Fields*, betwixt the *Windmill Meadow* and the road leading to *Chelfea* Bridge adjoining the gardener's ground. It flowers in *Auguft*.

It is mentioned, in the third edition of Mr. RAY's *Synopfis*, to have been found by Mr. SHERARD on the banks of fome watery pits beyond *Ely*, and by Mr. DALE about *Colchefter*. Mr. HUDSON defcribes it as growing plentifully about *Northfleet*; and Mr. LIGHTFOOT enumerates it among his *Scottifh* plants.

Some authors fufpect it to be poifonous. TRAGUS, in particular, mentions it as a plant fatal to fwine.

* Habet multa communia cum præcedente (viride) ut forte olim ab eodem ortum fit, quod racemorum ftructura indicat. *Fl. Suec.* n. 220. p. 80.

BUNIUM BULBOCASTANUM. EARTH-NUT.

BUNIUM *Lin. Gen. Pl.* PENTANDRIA DIGYNIA.

Corolla uniformis. *Umbella* conferta. *Fructus* ovatus.

Raii Syn. Gen. 11. UMBELLIFERÆ HERBÆ.

BUNIUM *Bulbocastanum. Lin. Syst. Vegetab.* p. 229. *Sp. Pl.* p. 349.

BULBOCASTANUM. *Haller Hist.* n. 783.

BULBOCASTANUM. *I. B.* III. 2. 30.

BULBOCASTANUM majus folio apii. *Bauhin. Pin.* 162.

BULBOCASTANUM majus et minus. *Ger. emac.* 1065.

NUCULA terrestris major. *Parkinf.* 893. *Raii Syn.* p. 209. Earth-nut, Kipper-nut, Pig-nut. *Hudson. Fl. Angl. ed.* 2. p. 122. *Lightfoot Fl. Scot.* p. 156. *Oeder. Fl. Dan. t.* 220.

RADIX perennis, tuberosa, extus castanei coloris, intus alba, solida, fibras tum ab imo tum a lateribus promens tenues, gustu subdulci grato, profunde in terra delitescente.

CAULIS pedalis, ad bipedalem, erectus, teres, firmulus, crassitie pennæ coracis, substriatus, glaber, penitus viridis, ramosus.

FOLIA radicalia longe petiolata, caulina sessilia, omnibus tenuissime divisis, fœniculi modo, saturate viridibus, marginibus foliorum ad lentem aculeato-ciliaris, *fig.* 7. *auct.* Spatha brevis, sulcatus, lævis, margine membranaceâ, albidâ.

UMBELLÆ plures, *universalis* multiplex radiis septem ad duodecim, *partialis* brevissima, conferta, radiis circiter duodecim.

INVOLUCRUM *universale* polyphyllum, lineare, breve, sæpe nullum; *partiale* setaceum, longitudine umbellulæ, aliquando nullum. *Perianthium proprium* vix manifestum,

COROLLA universalis uniformis, flosculi plerique fertiles; *propria* Petalis quinque, inflexo-cordatis æqualibus. *fig.* 2.

STAMINA: FILAMENTA quinque alba, subulata, corolla longiora, decidua. ANTHERÆ simplices, flavescentes. *fig,* 3.

PISTILLUM: GERMEN oblongum, inferum, viride. STYLI duo albi, suberecti. STIGMATA obtusi. *fig,* 4, 5.

SEMINA duo, oblongo-ovata, hinc convexa inde plana, glabra, subaromatica. *fig.* 6, 7.

ROOT perennial, tuberous, on the outside of a chesnut colour, within white, solid, putting forth slender fibres from the sides as well as the bottom, of an agreeable sweetish taste, lying deep in the ground.

STALK from one to two feet high, upright, round, stiffish, the thicknefs of a crow-quill, slightly striated, smooth, throughout of a green colour, and branched.

LEAVES from the root standing on long foot-stalks, those of the stalk sessile, all of them very finely divided like fennel, the small leaves edged with prickly hairs, *fig.* 1. *magn.* Sheath short, grooved, smooth, the edge membranous and whitish.

UMBELLS several, the *general* one composed of many rays from seven to twelve; the *partial* one very short, the rays about twelve and close.

INVOLUCRUM: the *general* one composed of many leaves, linear, short, often wanting; the *partial* one setaceous, the length of the small umbel, often wanting. *Partial Perianthium* scarcely manifest.

COROLLA: general Corolla uniform, most of the flowers fertile; individual one composed of five Petals, heart-shaped, bent in at top, and equal. *fig.* 2.

STAMINA: five FILAMENTS of a white colour, tapering, longer than the corolla, deciduous. ANTHERÆ simple and yellowish. *fig.* 3.

PISTILLUM: GERMEN oblong, beneath the corolla, of a green colour. STYLES two, white, nearly upright. STIGMATA blunt. *fig.* 4, 5.

SEEDS two, of an oblong-ovate shape, convex on one side and flat on the other, smooth, and somewhat aromatic. *fig.* 6, 7.

Children are frequently in the practice of digging up and eating the knobby roots of this plant, which, by some, are supposed to resemble the chesnut in its taste, whence its name of *Bulbocastanum.* Pigs also seek for it with avidity, from which circumstance it has also obtained the name of *Pig-nut.*

Few of our Umbelliferi have the characteristic marks which distinguish the Bunium; the principal of which are its knobbed root, and finely divided fennel-like leaves. The one which approaches the nearest to it is the *Oenanthe fistulosa,* especially when growing on ditch banks, where it is frequently thrown when the ditches are cleansed, or when it is cultivated in gardens; the roots, in such situations, are so similar to those of the Bunium, as to deceive even good judges. The radical leaves of the plant are also finely divided; it would be no wonder, therefore, if they should be mistaken for the Earth-nut.

A paper was published, about a year ago, in one of the Magazines, the *London,* if I mistake not, in which the roots of the *Oenanthe crocata,* well known for their poisonous effects, were said to have been eaten for those of this plant. We suspect, however, from various circumstances, that they were the roots of the *Oenanthe fistulosa.* It is our intention to make a more minute enquiry into this matter, and give our reasons more at large for this suspicion, when we figure that species.

The Earth-nut with us grows chiefly in woods, pastures, and orchards, and flowers in *June.*

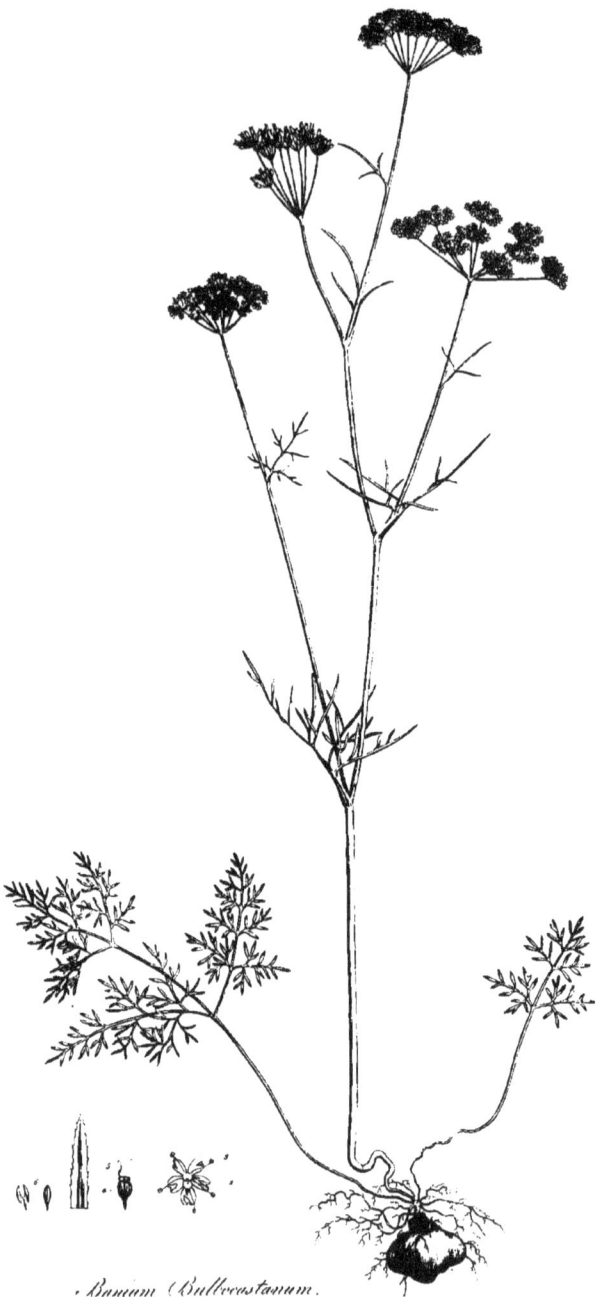

Bunium Bulbocastanum.

Chaerophyllum sylvestre.

CHÆROPHYLLUM SYLVESTRE. COMMON COW-PARSLEY.

CHÆROPHYLLUM. *Lin. Gen. Pl.* PENTANDRIA DIGYNIA.
Involucr. reflexum, concavum. *Petala* inflexo-córdata. *Fructut* oblongus, lævis.
Raii Syn. Gen. 11. UMBELLIFERÆ HERBÆ.
CHÆROPHYLLUM *sylvestre* caule lævi striato, geniculis tumidiusculis. *Lin. Syst. Vegetab.* p. 238.
Sp. Pl. p. 369. *Fl. Suec.* n. 257.
CEREFOLIUM foliis acute dentatis triplicato-pinnatis, glabris, nervis hirsutis. *Haller Hist.* n. 748.
CHÆROPHYLLUM *sylvestre.* *Scopoli Fl. Carn.* n. 366.
MYRRHIS sylvestris seminibus lævibus. *Bauhin. Pin.* 170.
CICUTARIA vulgaris. *Dod. Pempt.* 701.
MYRRHIS sylvestris. *Parkinson,* 935.
CICUTARIA alba Lugdunensis. *Ger. emac.* 1038. *Raii Syn.* p. 207. Wild Chervil. *Hudson. Fl. Angl.* p. 124. *Lightfoot Fl. Scot.* p. 167.

RADIX perennis, crassitie digiti intermedii, ad basin sæpius ramosa, extus pallide fusca, intus albida, succum vix lacteum fundens.

CAULIS bipedalis et ultra, erectus, crassitie digiti minimi, fistulosus, sulcatus, plerumque villosus, et sæpius purpurascens, ramosus, geniculatus, geniculis paululum incrassatis; rami suberecti, minus pubescentes.

FOLIA radicalia multiplicato-pinnata, sæpe amplissima; Costæ hirsutæ, fistulosæ, dorso acute angulatæ, antice canaliculatæ, orteæque ex vagina brevi striataque, pinnulæ lanceolatæ, acutæ, serratoiucisæ, plerumque hirsutæ; caulina et ramea successive minora minusque divisa, cæterum similia, superiori sæpe opposita aut terna cum totidem ramis axillaribus.

UMBELLÆ planiusculæ, nec densæ, nec plantæ ratione amplæ, ante anthesin nutantes, radiis compositæ a septem ad octodecim, gracilibus, teretibus et glabris.
INVOLUCRUM universale nullum, partiale pentaphyllum et ultra, foliolis acuminato-ovatis, concavis, glabris, ad oras subciliatis, umbellulis duplo brevioribus, tandem reflexis.

FLORES inodori, pauci steriles.
COROLLA: PETALA plana, et obverse ovata, albida, centralium flosculorum subæqualia, exteriorum vero extimum majus, et subcordato emarginatum. *fig.* 1, 2, 3.
STAMINA: FILAMENTA quinque albida, brevia et caduca. ANTHERÆ subrotundæ, didymæ, flavescentes. *fig.* 4.
PISTILLUM: GERMEN inferum, turbinatum, utrinque compressiusculum, lucidum; glandula nectarifera coronatum. STYLI duo, albi, parviusculi. STIGMATA simplicia. *fig.* 5, 6.

SEMINA duo, oblonga, antice sulcata, cæterum teretia, lævia, nitida, nigricantia, inodora, sapore vix ullo. *fig.* 7.

ROOT perennial, the thickness of the middle finger, most commonly branching out from the base; externally of a pale-brown colour, internally whitish, full of a juice which can scarcely be called milky.

STALK two feet high and upwards, upright, the thickness of the little finger, hollow, grooved, generally villous, and most commonly purplish, branched, jointed, joints somewhat thickened; branches nearly upright, less hoary than the stalk.

LEAVES next the root many times pinnated, often very large; the Ribs hirsute, fistulous, forming a sharp angle on the back, hollow in front, and arising from a short striated sheath, the small pinnæ lanceolate, pointed, deeply and irregularly serrated, generally hirsute, the leaves of the stalk and branches successively smaller, and less divided, in other respects similar, the upper ones often opposite or growing three together; with as many axillary branches.

UMBELLS flattish, neither thick nor large for the size of the plant, drooping before the expansion of the flowers, composed of seven to eighteen radii, which are slender, round, and smooth.
INVOLUCRUM the general one wanting, the partial one composed of five leaves or more, which are ovate, pointed, hollow, smooth, somewhat hairy on the edges, twice as short as the small umbells, finally turned back.

FLOWERS scentless, a few of them barren.
COROLLA: PETALS flat, inversely ovate, whitish, those of the central flowers nearly equal, but the outermost of the outer ones largest, somewhat heart-shaped and nicked. *fig.* 1, 2, 3.
STAMINA: FILAMENTS five, whitish, short, and deciduous. ANTHERÆ nearly round, double, and yellowish. *fig.* 4.
PISTILLUM: GERMEN placed beneath the corolla, broadest at top, flattish on both sides, shining, crowned with a nectariferous gland. STYLES two, white and small. STIGMATA simple. *fig.* 5, 6.

SEEDS two oblong, with a groove in the fore-part, round, smooth, shining, blackish, without scent, and almost tasteless. *fig.* 7.

In many parts of the kingdom this plant is well known by the name of *Cow-parsley*, a term we adopt in preference to *Cow-weed*, or *wild Chervil*; the former being applicable to the *Cow-parsnep* also, and the latter more properly belonging to the *Scandix Cerefolium* and *odorata*.

LINNÆUS's specific character of this *Chærophyllum* is only applicable to the upper part of the plant: the lower part of the stalk, by which it is most obviously distinguished, is strongly grooved, and covered with numerous short hairs.

It is one of the most common, as well as the earliest in blossom, of our *umbelliferi*, flowering in warm situations in *April*, and generally with us in *May*. It grows chiefly in orchards, pastures, and under hedges. In sheltered situations it produces a large crop of early foliage; hence it has been recommended by some writers on agriculture as worthy the attention of the Farmer, more especially as cows are said to be fond of it. To rabbits it is a luxurious treat, as those who keep them pretty generally know. In time of scarcity the young leaves, in some parts of the kingdom, are used as a pot-herb: the *boiled roots* are said to have a poisonous quality, perhaps without any foundation.

LINNÆUS remarks, in his *Flora Suecica*, that its presence indicates a fertile soil; and that its flowers are capable of communicating a yellow dye to woollen cloth.

Myosurus minimus.

MYOSURUS MINIMUS. MOUSE-TAIL.

MYOSURUS *Lin. Gen. Pl.* PENTANDRIA POLYGYNIA.

Cal. 5-phyllus, bafi adnatus. *Nectaria* 5 fubulatæ, petaliformia. *Sem.* numerofa.

Raii Syn. Gen. 15. HERBÆ SEMINE NUDO POLYSPERMÆ.

MYOSURUS *minimus. Lin. Syst. Vegetab.* p. 254. *Sp. Pl.* p. 407. *Fl. Suec.* n. 276.

MYOSURUS. *Haller. Hist.* n. 1159.

HOLOSTEO adfinis Cauda muris. *Bauh. Pin.* 190.

CAUDA MURINA. *Dod. Pempt.* 112.

RANUNCULUS gramineo folio, flore caudato, feminibus in capitulum fpicatum congeftis. *Tournefort Inst.* 293.

MYOSUROS. *Dillen. Nov. Gen.* p. 108. t. 4.

HOLOSTEUM Loniceri, Cauda muris vocatum. *Park.* 500.

CAUDA MURIS. *Gerard emac.* 426. *Raii Syn.* 251. Moufe-tail. *Lightfoot Fl. Scot.* p. 179. *Hudfon Fl. Angl.* ed. 2. p. 130.

RADIX annua, fibrofa.

FOLIA, omnia radicalia, viginti circiter in plantâ mediæ magnitudinis, erecta, longitudinê inæquali, linearia, fuperne latiora, compreffa, utrinque obfolete canaliculata, glabra, fubcarnofa, obtufa, e flavo-viridia, bafi rubicunda.

SCAPI quatuor, quinque, aut plures, uniflori, erecti, bipollicares et ultra, foliis longiores, teretes, fuperne paulo craffiores, glabri.

CALYX: PERIANTHIUM pentaphyllum, foliolis oblongis, obtufis, concavis, herbaceis, patentibus. *Unguibus* poftice elongatis, appreffis, acuminatis. *fig.* 1, 2.

COROLLA: PETALA quinque, calyce breviora, minutiffima, flavefcentia, unguiculata, limbo patente, bafi fubtubulofa. *fig.* 3. auct. 5.

STAMINA: FILAMENTA quinque, vel plura, longitudine fere calycis, receptaculo inferta, filiformia, alba, erecta. ANTHERÆ oblongæ, lutefcentes. *fig.* 4.

PISTILLUM: GERMINA numerofa, receptaculo in formam conico-oblongam infidentia. STYLI nulli. STIGMATA minima, fimplicia. *fig.* 6.

PERICARPIUM nullum. *Receptaculum* longiffimum, ftyliforme, feminibus imbricatim difpofitis tectum. *fig.* 7.

SEMINA numerofa, compreffa, mucronata. *fig.* 8, 9.

ROOT annual and fibrous.

LEAVES, all of them radical, about twenty in a plant of a middling fize, upright, of unequal length, linear, broadeft at top, and flattened, faintly channeled on each fide, fmooth, fomewhat flefhy, blunt, of a yellowifh green colour, and reddifh at the bafe.

FLOWERING STEMS four, five, or more, each fupporting one flower, upright, two inches or more in length, longer than the leaves, round, fmooth, and a little thickeft at the top.

CALYX: a PERIANTHIUM of five leaves, the leaves oblong, obtufe, hollow, herbaceous, fpreading. *Claws* lengthened out behind, preffed to the flowering ftem, and tapering to a point. *fig.* 1, 2.

COROLLA: five PETALS, fhorter than the calyx, very minute, of a yellowifh colour, clawed, the limb fpreading, fomewhat tubular at its bafe. *fig.* 3. magn. 5.

STAMINA: FILAMENTS five or more, almoft the length of the calyx, fixed to the receptacle, thread-fhaped, white, upright. ANTHERÆ oblong and yellowifh. *fig.* 4.

PISTILLUM: GERMINA numerous, fitting on a recepctacle, of an oblong, conic fhape. STYLES none. STIGMATA very fmall and fimple. *fig.* 6.

SEED-VESSEL none. *Receptacle* very long, ftyle-fhaped, covered with feeds, laying one over another. *fig.* 7.

SEEDS numerous, flat, and pointed. *fig.* 8, 9.

Some of the early Botanifts gave to this plant the name of *Moufe-tail*, more from the general appearance of the plant than any particular attention to generic character. TOURNEFORT afterwards arranged it with the *Ranunculi*. DILLENIUS, not fatisfied with fuch arrangement, made a diftinct genus of it; which has been adopted by LINNÆUS. The ftructure of the whole fructification is fingular, and deferving the attention of the young Botanift; in particular, he fhould be careful to diftinguifh the Corolla from the Stamina.

This delicate little annual grows in corn and clover-fields, and by path-fides, efpecially where water has been apt to ftagnate. It is found about *Iflington, Paddington,* and *Pancrafs*; but with more certainty in the fields about *Dulwich,* efpecially on the right-hand-fide of *Lordfhip-lane,* near *Dulwich Wood.*

It flowers in *May* and *June,* and ripens it feed in *July* and *Auguft.*

Peplis Portula.

PEPLIS PORTULA. WATER PURSLANE.

PEPLIS *Lin. Gen. Pl.* HEXANDRIA MONOGYNIA.

Perianth. campanulatum : ore 12-fido. *Petala* 6, calyci inferta. *Capf.* 2-locularis.

Raii Syſ. Gen. 25. HERBÆ HEXAPETALÆ ET POLYPETALÆ VASCULIFERÆ.

PEPLIS *Portula* floribus apetalis. *Lin. Syſt. Vegetab.* p. 283. *Sp. Pl.* p. 474. *Fl. Suec.* n. 311.

PEPLIS petalis ſenis, ſæpiſſime apetala. *Haller. Hiſt.* n. 856.

ALSINE paluſtris minor ſerpyllifolia. *Bauh. pin.* 120.

ANAGALLIS ſerpyllifolia aquatica. *I. B.* III. p. 372.

GLAUX aquatica folio ſubrotundo. *Loeſel.* p. 106. *ic.* 20.

GLAUX altera ſubrotundo folio. *Boccone* t. 84. *Vaillant. Bot. par.* t. 15. f. 5.

GLAUCOIDES paluſtre portulacæ folio, purpureo flore. *Michel.* p. 21. t. 18·

PORTULA. *Dillen. Nov. Pl. Gen.* p. 133. t. 7.

ALSINE rotundifolia ſ. Portulaca aquatica. *C.r. em.* 614.

ALSINE aquatica minor folio oblongo ſ. Portulaca aquatica. *Park.* 1260. *Raii Syn.* p. 368. Water Purſlane. *Hudſon. Fl. Angl. ed.* p. 147. *Lightfoot Fl. Scot.* p. 187.

RADIX annua, fibroſa.	ROOT annual and fibrous.
CAULIS palmaris et ultra, tetragonus, repens, ramoſus, lævis, rubicundus, geniculatus.	STALK a hand's breadth or more in length, ſquare, creeping, branched, ſmooth and red.
FOLIA oppoſita, ſubrotunda, petiolata, integerrima, glabra, nervo medio rubicundo.	LEAVES oppoſite, roundiſh, ſtanding on foot-ſtalks, perfectly entire, ſmooth and ſhining, the mid-rib reddiſh.
FLORES oppoſiti, axillares, ſeſſiles, minimi.	FLOWERS oppoſite, in the alæ of the leaves, ſeſſile and very ſmall.
CALYX: PERIANTHIUM monophyllum, campanula-tum, perſiſtens, maximum, ore duodecimfido, denticulis alternis minoribus, reflexis. *fig.* 1.	CALYX: a PERIANTHIUM of one leaf, bell-ſhaped, permanent, very large (in proportion to the petals), the mouth cut into five ſegments, of which the alternate ones are leaſt and turned back. *fig.* 1.
COROLLA: PETALA raro ſex, ovata, minutiſſima, roſea, calycis fauci inſerta. *fig.* 2.	COROLLA: PETALS ſeldom ſo many as ſix, ovate, very ſmall, roſe-coloured, inſerted into the mouth of the calyx. *fig.* 2.
STAMINA: FILAMENTA ſex, ſubulata, brevia. AN-THERÆ ſubrotundæ. *fig.* 3.	STAMINA: ſix FILAMENTS, tapering and ſhort. AN-THERÆ roundiſh. *fig.* 3.
PISTILLUM: GERMEN ovatum. STYLUS breviſſimus. STIGMA orbiculatum. *fig.* 4.	PISTILLUM: GERMEN ovate. STYLE very ſhort. STIGMA round. *fig.* 4.
PERICARPIUM: CAPSULA ſubrotunda, bilocularis, tenuiſſima, pellucida. *fig.* 5.	SEED-VESSEL: a roundiſh CAPSULE of two cavities, extremely thin and pellucid. *fig.* 5.
SEMINA plurima, minima, albida, angulata. *fig.* 6.	SEEDS numerous, very minute, whitiſh and angular.

The numerous ſynonyms prefixed to the deſcription of this plant ſhew in what a variety of lights it has been viewed by different Botaniſts. Thoſe of early times called it *Alſine*, as they did almoſt every plant whoſe parts of fructification were ſmall, and which bore the moſt diſtant ſimilitude to Chick-weed.

MICHELI and DILLENIUS in their reſpective *Nova Genera* wrought much reformation in theſe minute plants; and if they did not arrive at the ultimatum of deſcription and figure, they paved the way to it.

The *Peplis Portula* is a very common plant in pools of water that are apt to be dried up in the ſummer, particu-larly on heaths. It flowers from June to Auguſt. Its petals are very minute, and frequently fall ſhort of their proper number.

Polygonum amphibium.

POLYGONUM *Lin. Gen. Pl.* OCTANDRIA TRIGYNIA.

Cal. o. *Cor.* 5-partita, calycina. *Sem.* 1. angulatum.

Raii Syn. Gen. 5. HERBÆ FLORE IMPERFECTO SEU STAMINEO VEL APETALO POTIUS.

POLYGONUM *amphibium* floribus pentandris femidigynis, fpica ovata. *Lin. Syft. Vegetab.* p. 312. *Sp. Pl.* 517. *Fl. Suec.* n. 341.

POLYGONUM foliis ovato lanceolatis ciliatis, fpicis ovatis. *Haller Hift.* n. 1565.

POTAMOGETON falicis folio. *Bauh. pin.* 193.

FONTALIS major longifolia. *Parkinf.* 1254.

POTAMOGITON anguftifolium. *Ger. emac.* 821.

PERSICARIA falicis folio perennir. *Raii Syn.* Perennial Willow-leaved Arfmart, commonly called narrow-leaved Pondweed. *Hudfon Fl. Angl. ed.* 2. p. 169. *Lightfoot Fl. Scot.* p. 207. *Order Fl. Dan.* t. 282. *Threlkeld Syn. Stirp. Hibern.*

RADIX perennis, repens, craffitie culmi tritici majoris, e rubro-fufca, ex hortis five ægris difficillime eruta.	ROOT perennial, creeping, the thicknefs of a large wheaten ftraw, of a reddifh brown colour, with the greateft difficulty rooted out of gardens or fields.
CAULIS fefquipedalis et ultra, adfcendens, plerumque fimplex, teres, fiftulofus, fæpius ruber, geniculis tumidiufculis.	STALK a foot and a half or more in length, bending upward, generally fimple, round, hollow, moft commonly of a red colour, the joints a little fwelled.
FOLIA petiolata, cordato-lanceolata, hirfutula, ciliata, rigidula, faturate viridia, fæpe maculata, fubundulata, alterna, patentia.	LEAVES ftanding on footftalks, lanceolate, with a heart-fhaped bafe, flightly hirfute, edged with hairs, harfh to the touch, of a deep green colour, often fpotted, fomewhat waved, alternate, and fpreading.
STIPULÆ longæ, muticæ, hirfutæ, vaginantes, caulem arcte cingentes.	STIPULÆ long, not fringed with hairs at the extremity, hirfute, forming a fheath, which clofely furrounds the ftalk.
PEDUNCULI folitarii, feu gemini, hirfuti, rubri, ad unum latus compreffo-fulcati, vix terminales.	FLOWER-STALKS growing fingly, or two together, hirfute, red, on one fide flattened, and flightly grooved, not properly terminal.
SPICÆ ruberrimæ, primo pyramidales, demum ovatæ.	SPIKES of the flowers of a bright red colour, at firft pyramidal, afterwards ovate.
CALYX: PERIANTHIUM pedicellatum, quinque-partitum, coloratum, perfiftens, laciniis ovatis, obtufis. *fig.* 1.	CALYX: a PERIANTHIUM on a footftalk, divided into five fegments, coloured and permanent, the fegments ovate and obtufe. *fig.* 1.
COROLLA nulla.	COROLLA wanting.
STAMINA: FILAMENTA quinque, aut fex, fundo calycis inferta, fubulata, alba, calyce longiora; ANTHERÆ didymæ; POLLEN album, globofum. *fig.* 2.	STAMINA: five FILAMENTS inferted into the bottom of the calyx, tapering, white, longer than the calyx; ANTHERÆ double; POLLEN white and globular. *fig.* 2.
PISTILLUM: GERMEN fubovatum, rubrum; STYLUS ad medium ufque divifus; STIGMATA duo, rotunda. *fig.* 3.	PISTILLUM: GERMEN fomewhat ovate, and of a red colour; STYLE divided as far as the middle; STIGMATA two, round. *fig.* 3.
SEMEN ovatum, utrinque compreffum, acutum, nitidum, e nigro caftaneum. *fig.* 4.	SEED ovate, flattened on each fide, pointed, gloffy, of a dark chefnut colour. *fig.* 4.

The fpecies of *Polygonum* here figured acquires the name of *amphibium* from its growing both on land and in water; in ponds its leaves ufually float, whence the antient botanifts, regardlefs of its fructification, improperly called it a *Pondweed*, thus CASPAR BAUHINE gives it the name of *Potamogeton falicis folio*; RAY very properly corrects this error, and calls it *Perficaria Salicis folio perennis*.

Of the *Perficaria* divifion of the genus *Polygonum Linnæi*, this is the only one that has a perennial root, a character which at once diftinguifhes it as a fpecies, befides this it has many peculiarities, the leaves are heart-fhaped at the bafe, which with the edges fringed with hairs, they are alfo harfher to the touch than any of the other Perficaria's, efpecially when the plant grows out of the water, its ftipulæ more clofely furround the ftalk, which has generally two fpikes of flowers at its extremity, neither of which are perfectly terminal, thefe fpikes are more pyramidal when young, and of a brighter red colour than any of the fpecies related to it; when it grows in the water the whole plant becomes fmoother and is more difpofed to bloffom; botanifts have alfo obferved that the ftamina, which when the plant grows on land are longer, are here fhorter than the corolla*.

As a weed few plants are more pernicious, Batterfea Fields, in which this plant abounds, bear fufficient teftimony to the truth of this affertion, as its roots not only creep, but penetrate fo deep into the earth that they are feldom or never eradicated; in the drier part of the fields it feldom flowers, but in certain fpots, where the water has fettled in wet feafons, it flowers abundantly in September.

From its bloffoming thus rarely in arable land it fortunately is a more local plant than many of the fame genus, the feeds of which are either fown or introduced with manure.

To atone for its mifchievous effects on land, it contributes highly to ornament ponds, rivers, and pieces of water; thofe who wifh to cultivate it for this purpofe need only plant the roots near the water's edge, the plants will quickly find their way into the water, where they muft be fecured from water-fowl, who are fond of its feeds.

* This is by no means conftant.

Polygonum Convolvulus.

POLYGONUM CONVOLVULUS. CLIMBING BUCK-WHEAT.

POLYGONUM *Lin. Gen. Pl.* OCTANDRIA TRIGYNIA.

Cal. o. Cor. 5-partita, calycina. Sem. 1. angulatum.

Raii Syn. Gen. 5. HERBÆ FLORE IMPERFECTO SEU STAMINEO VEL APETALO POTIUS.

POLYGONUM *Convolvulus* foliis cordatis, caule volubili angulato, floribus obtusatis. *Lin. Syst. Vegetab.* p. 313. *Sp. Pl.* p. 522. *Fl. Suec.* n. 344.

POLYGONUM caule volubili, foliis sagittatis. *Haller. Hist.* n. 1561.

POLYGONUM *Convolvulus.* *Scopoli Fl. Carn.* n. 469.

CONVOLVULUS minor semine triangulo. *Bauh. Pin.* 295.

HELXINE semine triangulo. *IB.* II. 157.

VOLUBILIS nigra. *Ger. emac.* 863.

CONVOLVULUS minor Atriplicis folio. *Park.* 171.

FEGOPYRUM scandens sylvestre. *Raii Syn.* p. 144. Black Bind-weed. *Hudson Fl. Angl.* ed. 2. p. 171. *Lightfoot Fl. Scot.* 208.

RADIX annua, fibrosa, fusca.

ROOT annual, fibrous, of a brown colour.

CAULIS volubilis, tortus, ramosus, ramis alternis, fulcro destitutus, procumbens, pedalis, id nactus circa calamos et fruticulos scandit sæpe ad altitudinem usque humanam.

STALK twining, twisted, branched, branches alternate, when destitute of support, procumbent and about a foot in length, when growing about corn or shrubs often reaching the height of six feet.

FOLIA petiolata, sagittata, glabra, integerrima, inferne solitaria, superne bina et terna, inferioribus frequenter coloratis.

LEAVES standing on foot-stalks, arrow-shaped, smooth, perfectly entire, on the lower part of the stalk standing singly, on the upper part two and three together, the lowermost often coloured.

STIPULÆ parvæ, vaginantes, muticæ.

STIPULÆ small, forming a sheath round the stalk.

FLORES racemosi, pedunculati, in fasciculos pendulos dispositi.

FLOWERS growing in racemi, standing on foot-stalks, and disposed in pendulous clusters.

RACEMI longi, alterni.

FLOWER-BRANCHES long, and alternate.

CALYX: PERIANTHIUM quinquepartitum, persistens, laciniis tribus exterioribus majoribus, carinatis, viridibus, margine membranaceis, interioribus petaliformibus, coloratis. *fig.* 1.

CALYX: 1 PERIANTHIUM divided into five segments, and permanent, the three outermost segments larger, keeled, green and membranous on the edge, the innermost petal-like and coloured. *fig.* 1.

COROLLA nulla.

COROLLA none.

STAMINA: FILAMENTA octo, fundo calycis inserta, brevissima, subulata. ANTHERÆ purpureæ, didymæ. *fig.* 3.

STAMINA: eight FILAMENTS fixed into the bottom of the calyx, very short and tapering. ANTHERÆ purple, formed of two lobes. *fig.* 3.

PISTILLUM: GERMEN viride, triquetrum. STYLUS brevissimus, staminibus paulo brevior. STIGMA capitatum, trilobum. *fig.* 4.

PISTILLUM: GERMEN green, three-cornered. STYLE very short, not quite so long as the stamina. STIGMA forming a little head, composed of three lobes. *fig.* 4.

PERICARPIUM nullum, calyx semen continens.

SEED-VESSEL none, the calyx containing the seed.

SEMEN unicum, trigonum, nigerrimum. *fig.* 5.

SEED a single one, three-cornered, very black. *fig.* 5.

Some of the ancient Botanists, inattentive to the fructification, arrange this plant with the *Convolvuli* or *Bind-weeds.* RAY calls it *Fegopyrum scandens sylvestre,* but retains the old *English* name of Black Bind-weed. LINNÆUS classes it with the *Polygonum,* or *Knot-grass,* in which genus he also includes the *Bistorta,* the *Persicaria,* and the *Fagopyrum;* he could not, perhaps, scientifically have made more genera of them; yet nature, in all our *English* plants at least, keeps up a strong distinction between them, and as the old established names of *Bistort, Persicaria* or *Arsmart, Knot-grass,* and *Buck-wheat,* have no tendency in them to mislead, we have thought it better, in the present instance, to continue their use, than to adopt the new-fangled names of *Buck-wheat Knot-grass,* or *Binding Snake-weed.* In its fructification this plant is very similar to the Buck-wheat; but differs in having a twining stalk, with which it frequently twists round other plants, and proves injurious to them, especially in gardens and cultivated fields, where it often grows extremely rampant; in poor land it is an humble plant.

It flowers in *July* and *August.*

Its seeds afford excellent food for small birds. One year I observed its foliage, together with that of the Passion Flower, very much eaten by the *Ear-wig,* a well known enemy to certain flowers, and no less destructive to Caterpillars, and which, like the *Cock-roach,* is principally active under the veil of night.

Silene anglica.

SILENE ANGLICA. ENGLISH CATCHFLY.

SILENE *Lin. Gen. Pl.* DECANDRIA TRIGYNIA.

Cal. ventricofus. *Petala* 5-unguiculata : coronata ad faucem. *Capf.* 3-locularis.

Rail Syn. Gen. 24. HERBÆ PENTAPETALÆ VASCULIFERÆ.

SILENE *anglica* hirfuta petalis emarginatis, floribus erectis, fructibus reflexis pedunculatis alternis. *Lin. Syft. Vegetab.* p. 350. *Sp. Pl.* 594.

SILENE *anglica* hirfuta petalis fubintegerrimis, floribus fubfpicatis pedunculatis alternis erectis, fructibus divaricato-reflexis. *Hudfon Fl. Angl.* ed. 2. p. 187.

VISCAGO ceraftii foliis vafculis pendulis Anglica. *Dill. Elth.* 417. *t.* 309. *f.* 398.

LYCHNIS fylveftris hirfuta annua, flore minore albo. *Vaill. Parif.* 121. *t.* 16. *f.* 12.

LYCHNIS fylveftris flore albo minimo. *Raii Syn.* p. 339. Small Corn Campion, with a very fmall white flower.

RADIX annua, fimplex.

CAULIS plerumque pedalis, et ultra, erectus, teres, hirfutus, vifcidus, articulatus, geniculis tumidis, ramofus, in horto culta ramofior, debiliorque evadit, et geniculi magis intumefcunt.

RAMI alterni, cauli fimiles.

FOLIA oppofita, connata, lanceolata, fubundulata, integerrima, hirfutula, fubvifcida, punctis prominulis afperula, ad bafin pilis longis ciliata.

FLORES albidi, minimi, axillares, fubfpicati, pedunculati.

PEDUNCULI variæ longitudinis unguiculares et pollicares, teretes, fuperne incraffati, vifcidi, primo erecti, dein reflexi, demum feminibus maturis fuberecti.

CALYX: PERIANTHIUM monophyllum, tubulofum, quinquedentatum, dentibus apice rubris, oblongum, fulcis decem profunde exaratum, pilofum, vifcidum, perfiftens. *fig.* 1.

COROLLA: PETALA quinque. *Ungues* angufti, longitudine calycis. *Lamina* integra feu emarginata, lateribus fæpe involutis, fquamula ad bafin laminæ bifida, erecta. *fig.* 2.

STAMINA: FILAMENTA decem, fubulata, alba, ad lentem hirfuta. ANTHERÆ oblongæ, bilobæ, purpureæ. POLLEN album. *fig.* 3.

PISTILLUM: GERMEN viride, nitidum, fubconicum; STYLI tres, albi, erecti, villofi, germine breviores. *fig.* 4, 5. Glandula nectarifera ad bafin germinis. *fig.* 6.

PERICARPIUM: CAPSULA ovata, calyce tecta, trilocularis, apice fexfariam dehifcens. *fig.* 7.

SEMINA plurima, nigricantia, reniformia, ad lentem afpera. *fig.* 8.

ROOT annual and fimple.

STALK about a foot or more in height, upright, round, hirfute, vifcid, jointed, the joints fwelled, branched; cultivated in the garden, it becomes more branched, weaker, and the joints more fwelled.

BRANCHES alternate, like the ftalk.

LEAVES oppofite, connate, lanceolate, fomewhat waved, entire, flightly hairy, and fomewhat vifcid, roughifh, with little prominent points, at the bafe edged with longer hairs.

FLOWERS whitifh, very fmall, growing from the alæ of the leaves, forming a kind of fpike, ftanding on foot-ftalks.

FLOWER-STALKS of various lengths, from half an inch to an inch, round, thickened upwards, vifcid, at firft upright, afterwards turned downwards, finally, when the feeds are upright, becoming nearly upright.

CALYX: a PERIANTHIUM of one leaf, tubular, having five teeth, which are red at the tips, oblong, marked with ten deep grooves, hairy, vifcid, and permanent. *fig.* 1.

COROLLA: five PETALS. *Claws* narrow, the length of the calyx. *Lamina* entire, or nicked, the fides often rolled in, the fcale at the bafe of the lamina bifid and upright. *fig.* 2.

STAMINA: ten FILAMENTS, tapering, white, hairy when magnified. ANTHERÆ oblong, formed of two lobes and purple. POLLEN white. *fig.* 3.

PISTILLUM: GERMEN green, fhining, fomewhat conical. STYLES three, white, upright, villous, fhorter than the germen. *fig.* 4, 5. A nectariferous Gland at the bafe of the germen. *fig.* 6.

SEED-VESSEL: an ovate CAPSULE, covered by the calyx, of three cavities, opening at top, with fix teeth. *fig.* 7.

SEEDS numerous, blackifh, kidney-fhaped, rough when viewed with a magnifier. *fig.* 8.

The prefent, though not a fhewy plant, may be numbered among the more rare ones in the neighbourhood of London, as well as in many other parts of *Great Britain*; nor does it appear to be common throughout *Europe*: yet, in particular fpots, it is found in great plenty, as in the corn-fields about *Coombe Wood*. I have alfo feen it growing in great abundance in the corn-fields near *Newport*, in the *Ifle of Wight*.

It flowers in *July*.

DILLENIUS gives a figure of it in his *Hortus Elthamenfis*, which is too diminutive: his character of *vafcula pendula* is not too much to be depended on, fince it frequently happens that they are not fo.

Arenaria trinervia.

ARENARIA TRINERVIA. PLANTAIN-LEAVED CHICKWEED.

ARENARIA *Lin. Gen. Pl.* DECANDRIA TRIGYNIA.

Cal. 5-phyllus, patens. *Petala* 5, integra. *Capf.* 1 locularis, polysperma.

Rail Syn. Gen. 24. HERBÆ PENTAPETALÆ VASCULIFERÆ.

ARENARIA *trinervia* foliis ovatis acutis petiolatis nervosis. *Lin. Syst. Vegetab.* p 353. *Sp. Pl.* p. 605. *Fl. Suec.* n. 397.

ALSINE foliis ovato-lanceolatis, trinerviis. *Haller. Hist.* n. 878.

ALSINE *Plantaginis folio. I. B.* III. 364. *Raii Syn.* p. 349. Plantain-leaved Chickweed. *Hudson Fl. Angl. ed.* 2. p. 191. *Lightfoot Fl. Scot.* p. 230. *Oeder Fl. Dan.* t. 429.

RADIX annua, tenuissima, fibrosa, albida.

ROOT annual, very slender, fibrous, and whitish.

CAULES plures, spithamæi, pedales, et ultra, debiles, teretes, undique pubescentes, geniculati, ramosissimi.

STALKS several, a span, a foot or more in length, weak, downy, all round jointed, and very much branched.

FOLIA opposita, ovata, acuta, pallide viridia, trinervia, integerrima, margine nervisque minutim ciliatis, supremis sessilibus, inferioribus petiolatis, crebrioribus, minoribus.

LEAVES opposite, ovate, pointed, of a pale green colour, three-rib'd, entire, the edge and ribs finely fringed with hairs, the uppermost sessile, the lowermost standing on foot-stalks, more numerous and smaller.

PETIOLI subalati, marginibus pilosis.

LEAF STALKS somewhat winged, the edges hairy.

FLORES alterni, solitarii, e dichotomia caulis.

FLOWERS alternate, solitary, proceeding from the forking of the stalk.

PEDUNCULI teretes, pubescentes, primo erecti, demum horizontaliter extensi, apice subinflexo, et paululum incrassato.

FLOWER-STALKS round, downy, at first erect, finally horizontally extended, the tip somewhat bent in, and a little thickened.

CALYX: PERIANTHIUM pentaphyllum, foliis ovatoacuminatis, carinatis, pubescentibus, corolla longioribus. *fig.* 1.

CALYX: a PERIANTHIUM of five leaves, which are ovate, running out to a point, keeled, downy, longer than the corolla. *fig.* 1.

COROLLA: PETALA quinque, parva, alba, obovata, integra. *fig.* 2.

COROLLA: five PETALS, small, white, inversely ovate and entire. *fig.* 2.

STAMINA: FILAMENTA decem, longitudine corollæ, alba, filiformia. ANTHERÆ minutæ, flavæ. *fig.* 3.

STAMINA: ten FILAMENTS, the length of the corolla, white, filiform. ANTHERÆ very small, and yellow. *fig.* 3.

PISTILLUM: GERMEN ovatum. STYLI tres, longitudine germinis. STIGMATA obtusiuscula. *fig.* 4.

PISTILLUM: GERMEN ovate. STYLES three, the length of the germen. STIGMATA bluntish. *fig.* 4.

PERICARPIUM: *Capsula* subconica, tecta, unilocularis.

SEED-VESSEL: a *Capsule* of a shape somewhat conic, covered by the calyx, and having one cavity.

SEMINA plurima, subreniformia, planiuscula, *glaberrima*, nigra.

SEEDS numerous, somewhat kidney-shaped, flattish, very smooth, and black.

There exists a considerable similarity betwixt the present plant and the common Chickweed; the attentive observer will, however, find them to differ very materially.

As a principal part of the professed design of this work is to remove, as much as possible, every difficulty attending an investigation of the British plants, we shall point out those differences which have appeared to us the most striking in comparing the two together. The common Chickweed, as its name imports, is found almost every where, as well in exposed as in shady situations; this, on the contrary, is found with us only in woods, and on the shady banks surrounding them, and, compared with the other, may be considered rather as a scarce plant. The common Chickweed flowers in *March* and *April*; this produces its blossoms in *May* and *June*. The common Chickweed has a row of hairs running down each side of the stalk; this is uniformly covered with very short hairs, scarcely discernible. The former has a procumbent stalk; this grows erect. In the former the leaves are not distinguished by any veins or ribs; this, on the contrary, has three strong ones, which give them somewhat the appearance of those of Plantain, whence its name. In the former the petals are bifid; in this they are entire. The seeds also afford another very striking difference: in the common Chickweed they are brown and rough; while those of the *trinervia* are black, perfectly smooth, and shining.

We know of no particular use to which this diminutive plant is applicable.

ARENARIA SERPYLLIFOLIA. THYME-LEAVED CHICKWEED.

ARENARIA *Lin. Gen. Pl.* DECANDRIA TRIGYNIA.

Cal. 5-phyllus, patens. *Petala* 5 integra. *Capf.* 1-locularis, polyfperma.

Raii Syn. Gen. 24. HERBÆ PENTAPETALÆ VASCULIFERÆ.

ARENARIA *ferpyllifolia* foliis fubovatis acutis feffilibus, corollis calyce brevioribus. *Lin. Syft. Vegetab.* p. 353. *Sp. Pl.* p. 606. *Fl. Suec.* n. 398.

STELLARIA *ferpyllifolia. Scopoli. Fl. Carn.* n. 544.

ALSINE foliis ovato-lanceolatis, fubhirfutis, petalis calyce brevioribus. *Haller Hift.* n. 875.

ALSINE minor multicaulis. *Bauh Pin.* 250.

ALSINE minima. *Ger. emac.* 612. *Raii Syn.* 349. The leaft Chick-weed. *Hudfon. Fl. Angl.* p. 191. *Lightfoot Fl. Scot.* p. 230.

RADIX annua, fibrofa, albida.

ROOT annual, fibrous, and whitifh.

CAULES plerumque plures, palmares, fuberecti, teretes, pubefcentes, geniculati, ut plurimum fimplices, apice dichotomi, ramis bifidis.

STALKS for the moft part numerous, about four inches high, nearly upright, round, downy, jointed, for the moft part fimple, dichotomous at top, branches bifid.

FOLIA oppofita, ovata, acuta, feffilia, rigidula, hirfutula, ad lentem ciliata, nervo medio fubtus confpicuo, inferioribus crebrioribus.

LEAVES oppofite, ovate, pointed, feffile, a little rigid and flightly hirfute, vifibly ciliated when magnified, the mid-rib confpicuous on the under fide, the lowermoft leaves growing thickeft together.

FLORES albi, pedunculati.

FLOWERS white, ftanding on foot-ftalks.

PEDUNCULI teretes, foliis longiores, uniflori, e dichotomia caulis, ad unum latus inclinati, fuberecti.

FLOWER-STALKS round, longer than the leaves, fupporting one flower, proceeding from the forking of the ftalk, inclined to one fide, and nearly upright.

CALYX: PERIANTHIUM pentaphyllum, foliolis lanceolatis, acuminatis, hirfutulis, carinatis. *fig.* 1.

CALYX: a PERIANTHIUM of five leaves, which are lanceolate, tapering to a point, fomewhat hairy and keeled. *fig.* 1.

COROLLA: PETALA quinque, alba, ovata, obtufa, calyce duplo fere breviora. *fig.* 2.

COROLLA: five petals, of a white colour, ovate, obtufe, about half the length of the calyx. *fig.* 2.

STAMINA: FILAMENTA decem alba, capillaria, longitudine corollæ. ANTHERÆ fubrotundæ, albæ, *fig.* 3.

STAMINA: ten FILAMENTS, white, very fine, the length of the corolla. ANTHERÆ roundifh and white. *fig.* 3.

PISTILLUM: GERMEN viride, fubrotundum. STYLI tres, albi, filiformes reflexi. STIGMATA fimplicia. *fig.* 4.

PISTILLUM: GERMEN green, roundifh. STYLES three, white, filiform and reflexed. STIGMATA fimple. *fig.* 4.

PERICARPIUM: *Capfula* ovata, fubventricofa, tecta, unilocularis, apice quinquefariam feu fex fariam dehifcens. *fig.* 5.

SEED-VESSEL: an ovate *Capfule*, fomewhat bellying out at bottom, covered by the calyx, of one cavity, opening at top, with five or fix teeth. *fig.* 5.

SEMINA plurima, minima, reniformia, ad lentem lineis infculptis pulchre reticulata. *fig.* 6.

SEEDS numerous, very fmall, kidney-fhaped, beautifully reticulated with impreffed lines, vifible when magnified. *fig.* 6.

This plant, one of the leaft of the genus *Arenaria*, is very common on walls, among rubbifh, and in dry and barren places. It flowers in *June*.

There is a neatnefs in it fufficient to recommend it as an ornamental plant for walls, rocks, &c. on which it will grow moft readily.

The rigidity of its ftalks, and thyme-like form of its leaves, readily diftinguifh it from all its congeners.

Arenaria serpyllifolia.

Sedum saxangulare ?

Sedum sexangulare. insipid Stonecrop.

SEDUM *Lin. Gen. Pl.* Decandria Pentagynia.

> *Cal.* 5 fidus. *Cor.* 5 petala. *Squamæ* nectariferæ 5 ad bafin germinis. *Caps.* 5.

Raii Syn. Gen. 17. Herbæ multisiliquæ seu corniculatæ.

SEDUM *fexangulare* foliis fubovatis adnato feffilibus gibbis erectiufculis fexfariam imbricatis. *Lin. Syft. Vegetab.* p. 359. *Spec. Plant.* p. 620. *Fl. Suecic. n* 404.

SEDUM foliis teretibus, ternatis; caulibus fimplicibus trifidis. *Haller. hift. n.* 965.

SEDUM *fexangulare. Scopoli Fl. Carn. n.* 558.

SEMPERVIVUM minus vermiculatum infipidum. *Baubin. pin.* 284.

SEDUM minimum luteum non acre. *Baubin. hift.* 3. p. 695. *Hudfon Fl. Angl. ed.* 1. p. 172.

RADIX perennis, fibrofa.

CAULES bafi repentes, floriferi erecti, tripollicares et ultra, teretes, glabri, punctati, inferne nudi, rubentes.

FOLIA oblonga, carnofa, teretiufcula, obtufa, erecto-patentia, fexfariam imbricata, præfertim ante florefcentiam, rigidula, adnato-feffilia, inferne rubentia, fuperne caulibus faltem floriferis e flavo viridia, infipida.

CYMA plerumque trifida, floribus in fingulo ramulo tribus ad quinque, feffilibus.

CALYX: Perianthium quinquepartitum, laciniis obtufis, carnofis, bafi tenuioribus.

COROLLA: Petala quinque, flava, lanceolata, acuminata, calyce duplo longiora, patentia. *fig.* 2.

STAMINA: Filamenta decem, fubulata, longitudine corollæ; Antheræ fubrotundæ, flavefcentes. *fig.* 3. 4.

PISTILLUM: Germina quinque, erecta, oblonga, defmentia in Stylos tenuiores; Stigmata fimplicia. *fig.* 5.

ROOT perennial and fibrous.

STALKS creeping at the bafe, thofe which produce flowers about three inches or more in height, round, fmooth, dotted, below naked and of a reddifh colour.

LEAVES oblong, flefhy, roundifh, obtufe, upright, but bending a little outward, placed one over another in fix rows, efpecially before he bloffoms open, fomewhat rigid, feffile, as if ftuck to the ftalk, thofe on the lower part of the ftem of a reddifh colour, on the upper part yellowifh, at leaft on the flowering ftalks, infipid.

CYMA generally divided into three branches, on each of which are placed from three to five flowers, without footftalks.

CALYX: a Perianthium divided into five fegments, which are obtufe, flefhy, and flenderer at the bafe.

COROLLA: five yellow Petals, lance-fhaped, acuminated, fpreading, twice the length of the calyx. *fig.* 2.

STAMINA: ten Filaments, tapering, the length of the corolla; Antheræ roundifh, and of a yellowifh colour. *fig.* 3. 4.

PISTILLUM: Germina five, upright, oblong, terminating in flender Styles: Stigmata fimple. *fig.* 5.

In Dillenius's edition of *Ray's Synopfis* this plant is omitted, and not confidered even as a variety of the *Sedum acre*. Mr. Hudson, in the fift edition of his *Flora Anglica*, introduced it as a diftinct fpecies, in which he followed the opinion of Linnæus; in his laft edition, without affigning any reafon, he makes it a variety of the *Sedum acre*; Haller, however, and Scopoli confirm Linnæus's opinion, and on fuch authority we furely may differ from Mr. Hudson.

The conftant want of that biting tafte which gives the name of *Wall Pepper* to the *Sedum acre*, has been confidered by many of our Englifh Botanifts fufficient to conftitute this a diftinct fpecies; for though acrid plants may fometimes become mild, as in the *Hydropiper*, yet inftances of that kind very rarely occur, but it is not in its tafte alone that the *fexangulare* differs from the *acre*, in its leaves we fhall find a fatisfactory difference, on comparing thefe together as they grow on the flowering ftems of both plants, we find thofe of the *acre* fhort, broad at the bafe, and at a confiderable diftance afunder, *vid. fig.* 1. while thofe of the *fexangulare* are nearly of the fame thicknefs throughout, longer, and more numerous, *vid. fig.* 1. we may alfo add, that they are in general much redder, in the young fhoots of the *fexangulare* the leaves form fix rows or angles, which are fometimes ftraight and fometimes oblique; no traces of which are vifible in the *acre*; another circumftance which adds fome weight to the foregoing is, that the *acre* flowers a fortnight fooner than the *fexangulare*; the parts of the fructification afford little or no difference, indeed a great famenefs in this refpect runs through the whole genus.

We find this plant growing plentifully on Greenwich-park-wall, the fouth fide, near the weftern corner.

It flowers about the latter end of June.

Spergula nodosa. Knotted Spurrey.

SPERGULA *Lin. Gen. Pl.* Decandria Pentagynia.

Cal. 5-phyllus. *Petala* 5 integra. *Capf.* ovata, 1-locularis, 5-valvis.

Raii Syn. Gen. 24. Herbæ Pentapetalæ Vasculiferæ.

SPERGULA *nodofa* foliis oppofitis fubulatis lævibus caulibus fimplicibus. *Lin. Syft. Vegetab.* p. 363. *Sp. Pl.* p. 630.

ALSINE foliis fuperioribus fafciculatis. *Haller. Hift.* n. 871.

STELLARIA *nodofa. Scopoli Fl. Carn.* n. 545.

ALSINE nodofa Germanica. *Bauh. Pin.* p. 251.

ALSINE paluftris, ericæ folio, polygonoides, articulis crebioribus, flore albo pulchello. *Pluk. alm.* 23. *t.* 7. *fig.* 4.

SAXIFRAGA paluftris Anglica. *Park.* 427.
ALSINE paluftris foliis tenuiffimis, feu Saxifraga paluftris Anglica. *Ger. emac.* 567. 568. *Raii Syn.* p. 350. Englifh Marfh-Saxifrage. *Hudfon Fl. Angl. ed.* 2. p. 203. *Lightfoot Fl. Scot.* p. 244.

RADIX perennis, fibrofa.

CAULES ex una radice plures, palmares et ultra, nunc procumbentes, nunc afcendentes, poft florefcentiam fæpe repentes, fimplices feu ramofi, teretes, tenues, glabri, parce pilofi, pilis ad lentem globuliferis, crebris geniculis intercepti, geniculis tumidis.

FOLIA *radicalia* plurima, cæfpitofa, læte virentia, linearia, acuta, uncialia, fubcarnofa, glabra, caulina inferiora paulo breviora, connata, fuperiora breviffima, teretiufcula, fafciculata, ex alis prolifera.

FLORES albi, delicatuli, in fummis caulibus et ramulis, majores quam pro plantulæ modo.

PEDUNCULI erecti, femipollicares.

CALYX: Perianthium pentaphyllum, foliolis oblongis, concavis, fubpilofis, pilis ut in caule. *fig.* 1.
COROLLA: Petala quinque, alba, calyce duplo longiora, ovato-rotundata, integerrima. *fig.* 2.

STAMINA: Filamenta decem, fubulata, alba, corollâ breviora. Antheræ concolores, incumbentes, primo bilobæ, lobis oblongis, parallelis. *fig.* 3.

PISTILLUM: Germen turbinatum. Styli quinque, filiformes, villofuli, reflexi. Stigmata fimpliciis, *fig.* 4.
PERICARPIUM: Capsula parva, ovata, calyce tecta, unilocularis, quinquevalvis.
SEMINA plurima, minima, nigricantia.

‡ ROOT perennial and fibrous.

STALKS feveral from one root, four inches or more in length, fometimes procumbent, fometimes nearly upright, after flowering often creeping, fimple or branched, round, flender, fmooth, fparingly haired, the hairs appearing globular at top when magnified, having numerous joints which are fwelled.

LEAVES next the root numerous, forming a turf, of a beautiful dark green colour, linear, pointed, about an inch in length, fomewhat flefhy, fmooth, the lowermoft ftalk-leaves a little longer than the radical ones, joined together at bottom, the uppermoft ftalk-leaves very fhort, from their alæ producing fmall tufts of leaves, the rudiments of branches.

FLOWERS white, and delicate, fitting on the tops of the ftalks and branches, large in proportion to the fize of the plant.

FLOWER-STALKS upright, about half an inch in length.

CALYX: Perianthium compofed of five leaves, which are oblong, hollow, flightly hairy, the hairs like thofe on the ftalk. *fig.* 1.
COROLLA compofed of five white petals, twice the length of the calyx, of a roundifh egg-fhape, perfectly entire. *fig.* 2.

STAMINA: ten Filaments tapering, white, fhorter than the corolla. Antheræ of the fame colour, lying acrofs the filament, at firft compofed of two oblong lobes parallel to each other. *fig.* 3.

PISTILLUM: Germen broad at bottom, narrow at top. Styles five, filiform, flightly villous and reflexed. Stigmata fimple. *fig.* 4.

SEED-VESSEL: a fmall ovate Capsule covered with the calyx, of one cavity and five valves.
SEEDS numerous, very minute, of a blackifh colour.

The *Spergula nodofa* recommends itfelf to our notice by the beauty of its verdure, and the delicacy of its bloffoms; the largenefs and whitenefs of which, joined to its place of growth, ferve alfo to diftinguifh it from thofe plants which may have fome refemblance to it in their foliage.

It grows in moift fituations, frequently among herbage, and fometimes out of Walls, Rocks, and Stones.

I have obferved it growing out of the wall by the *Thames* fide, in feveral places betwixt *Lambeth* and *Putney*.

I have alfo found it on *Hounflow Heath* with the *Sagina procumbens* and *Centunculus minimus*. It flowers in *July* and *Auguft*.

About *London* it is a fcarce plant; but in the north of England it is very common on the borders of rivulets, and grows generally more upright than with us: a fmall fpecimen of it, in this ftate, is reprefented on the plate.

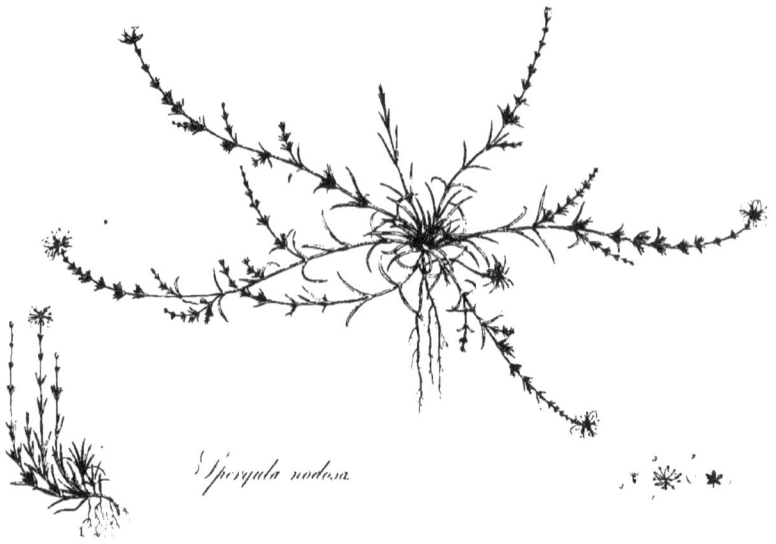

Spergula nodosa.

Spergula saginoides.

SPERGULA *Lin. Gen. Pl.* DECANDRIA PENTAGYNIA.

Cal. 5 phyllus. *Petala* 5, integra. *Capf.* ovata, 1-locularis, 5-valvis.

Raii Syn. Gen. 24. HERBÆ PENTAPETALÆ VASCULIFERÆ.

SPERGULA *faginoides* foliis oppofitis linearibus lævibus, pedunculis folitariis longiſſimis, caule repente. *Lin. Syſt. Vegetab.* p. 363. *Sp. Pl.* 631.

ALSINE tenuifolia pediculis florum longiſſimis. *Vaillant Botan. Paris,* p. 8. a. 11.

SPERGULA *laricina* foliis oppofitis fubulatis ciliatis fafciculatis, floribus pentandris. *Hudfon Fl. Angl. ed.* 2. p. 203.

SPERGULA *laricina. Lightfoot Fl. Scot.* p. 224.

SAXIFRAGA graminea pufilla foliis brevioribus craſſioribus et fucculentioribus. *Raii Syn.* p. 345?

RADIX perennis, fibrofa.

CAULES ex una radice plures, pollicares aut bipollicares, bafi procumbentes, et ut plurimum repentes, ramoſi, teretes, pilis brevibus glanduliferis vix confpicuis veſtiti.

FOLIA radicalia femipollicaria, linearia, acuta, mucrone albido terminata, faturate viridia, glabra, fubcarnofa, fuperne nuda, inferne et ad oras pilofa, pilis glanduliferis; caulina breviora, connata, planiufcula, fubfecunda. *fig.* 1.

PEDUNCULI fuberecti, fimplices, pollicares et ultra, teretes, ex fufco purpurafcentes, apice nutantes.

FLORES albi, pulchelli.

CALYX: PERIANTHIUM pentaphyllum, perfiftens, foliolis ovato-oblongis, concavis, obtufis, vifcidulis, margine membranaceis. *fig.* 2.

COROLLA: PETALA quinque, alba, longitudine calycis, fubrotunda, integerrima, patentia. *fig.* 3.

STAMINA: FILAMENTA quinque ad decem, fæpius vero quinque, fubulata, longitudine germinis; ANTHERÆ parvæ, luteæ. *fig.* 4.

PISTILLUM: GERMEN obovatum; STYLI plerumque quinque, filiformia, villofa, reflexa; STIGMATA fimplicia. *fig.* 5.

PERICARPIUM: CAPSULA uni-locularis, quinque-valvis, calyci infidens.

SEMINA plurima, minima, fufca, ad lentem punctata. *fig.* 7, 8.

ROOT perennial and fibrous.

STALKS, feveral arife from one root, an inch or two in length, procumbent, and generally creeping at bottom, branched, round, covered with fhort glandular hairs, fcarcely vifible.

LEAVES next the root about half an inch in length, linear, fharp and terminated by a whitifh point or briftle, of a deep green colour, fomewhat fhining, and rather flefhy, on the upper fide fmooth, on the under fide, and at the edge, hairy, the hairs terminated by little glands, thofe of the ftalk fhorter, growing together at the bafe, flattifh, and tending fomewhat one way. *fig.* 1.

FLOWER-STALKS fomewhat upright, fimple, an inch or more in length, round, of a brownifh purple colour, nodding at top.

FLOWERS white and pretty.

CALYX: a PERIANTHIUM of five leaves, permanent, the leaves oval, hollow, obtufe, fomewhat vifcid, the edge membranous. *fig.* 2.

COROLLA: five white PETALS, the length of the calyx, of a roundifh fhape, entire at the edge and fpreading. *fig.* 3.

STAMINA: FILAMENTS from five to ten, but moſt commonly five, tapering, the length of the germen; ANTHERÆ fmall and yellow. *fig.* 4.

PISTILLUM: GERMEN inverfely ovate; STYLES generally five, thread-fhaped, villous and turned back; STIGMATA fimple. *fig.* 5.

SEED-VESSEL: a CAPSULE of one cavity, and three valves fitting on the calyx. *fig.* 6.

SEEDS numerous, very fmall and brown, appearing dotted when magnified. *fig.* 7, 8.

We are led to confider this plant as the *Spergula faginoides* of LINNÆUS, from its according exactly with a minute defcription given of it by VAILLANT in his *Botanicon Parifienfe* *, to which the former refers, and furely no name was ever more aptly applied, for did not its confpicuous petals proclaim it, it might for ever have paffed for the *Sagina procumbens*; thefe lead us to the plant, and examination proves it to be a *Spergula*, inconftant indeed like many other plants in the number of its ftamina.

In its generic character this fpecies of *Spergula* differs in number only from the *Sagina procumbens*, it agrees particularly with it in the form of its capfules, fize, and fhape of its feeds, but two obvious fpecific characters at once diftinguifh them, thefe are the fize of the petals, and the peculiar hairinefs of the whole plant; in the *Sagina procumbens* the petals are very fmall, being much fhorter than the calyx, whence they are inconfpicuous; in the *Spergula* they are of the fame length as the calyx, and, when expanded, become very confpicuous; the *Sagina procumbens* (which muft not be confounded with the *apetala* LINNÆI) is fmooth, while the *Spergula* has its ftalks, leaves, peduncles and calyx covered with fhort hairs, having little globules at their extremities, and which are very diftinguifhable when magnified.

It is not uncommon on *Putney Heath*, and in fimilar fituations about *Coomb Wood, Surry*; Dr. GOODENOUGH difcovered it plentifully on *Bagfhot Heath*, efpecially on fome banks thrown up on *Potnell Warren*, near the great Bog at *Virginia Water*; Mr. LIGHTFOOT fhewed it me feveral years ago on *Uxbridge Moor*; and Mr. HUDSON mentions it as growing about *Cobham* and *Efher* in *Surry*. It flowers from June to Auguft.

C

* Cette plante a le port & les feuilles de l' *Alfine minima flore fugaci.* J. R. H. mais elle s'eleve plus haut. Ses tiges et les pedicules des fleurs font ordinairement brunes. Sa fleur n'a qu' environ 2 lignes de diametre. Elle eſt a 5 petales blancs, entiers ronds, qui ne debordent point le calice & qui font oppofé a fes cantons. Le piftile eft un petit bouton tirant fur l'ovale, vert pâle, furmonté de 5 ftiles blancs, courts diſpoſés en etoile, & entouré de 10 Etamines blanches ainfy que leurs fommets. Ces Etamines n'ont pas une ligne de long. Le calice eſt parfemé de petits poils tres courts. Il eſt decoupé en crois a 5 parties egales. Cette plante ne s'eleve que depuis 2 jufqu'à 4 pouces, elle pouffe ordinairement plufieurs tiges de fa racine, lefquelles fe couchent d'abord fur la terre, & font droites dans le reft de leur longueur. Ses feuilles font liffes, vertes, roides, dures et refemblent affes bien a celles du Knawel ou de l' *Alfine minima flore fugaci.* Elle commence a fleurir vers la fin de May & continue en Juin & Juillet. Elle fe trouve dans les fichus qui font au de la St. Leger entre la forêt et le Village de St. Lucien le long du chemin. Elle n'a que le gouft d'herbe. Son fruit s'ouvre ordinairement en 4 & quelquefois 5 lobes de la pointe vers la bafe & contient dans fa cavité plufieurs femences contrafre tres menues. *Botan. Par.* p. 8, 9.

Euphorbia exigua.

EUPHORBIA *Lin. Gen. Pl.* DODECANDRIA TRIGYNIA.

Cor. 4 f. 5-petala, calyci infidens. *Cal.* 1-phyllus, ventricofus. *Capf.* 3-cocca.

Raii Syn. Gen. 22. HERBÆ VASCULIFERÆ, FLORE TETRAPETALO ANOMALÆ.

EUPHORBIA *exigua* umbella trifida : dichotoma : involucellis lanceolatis, foliis linearibus. *Lin. Syf. Vegetab.* p. 375. *Sf. Pl.* p. 654.

TITHYMALUS foliis linearibus, ftipulis lanceolatis ariftatis. *Haller Hift.* n. 1048.

TITHYMALUS five Efula exigua. *Bauh. Pin.* p. 295.

ESULA exigua Tragi. *Ger. emac.* 502.

TITHYMALUS leptophyllus. *Parkinf.* 193. *Raii Syn.* 313. Dwarf Spurge, or fmall annual Spurge. *Hudfon Fl. Angl. ed.* 2. p. 208. *Lightfoot Fl. Scot.* p. 250.

RADIX annua, fimplex, paucis fibrillis inftructa.
CAULIS erectus, ramofus, foliofiffimus, femipedalis.

ROOT annual, fimple, furnifhed with few fibres.
STALK upright, branched, very leafy, about fix inches high.

RAMI plerumque inferiorem partem caulis tantummodo occupant, oppofiti, fuberecti.
FOLIA plurima, apprefla, linearia, obtufiufcula.

BRANCHES generally occupy the lower part of the ftalk only, are oppofite and nearly upright.
LEAVES numerous, prefled to the ftalk, linear, and fomewhat obtufe.

UMBELLA trifida, interdum quadrifida, rarius quinque-fida, dichotoma.
STIPULÆ *Umbellæ* lanceolato-lineares.

UMBELL dividing into three branches, fometimes four, rarely five, thofe forked.
STIPULÆ of the *general Umbell* of a fhape betwixt lanceolate and linear.

—— *Umbellulæ* ovato-oblongæ, acuminatæ, oppo-fitæ, fæpe inæquales.

—— of the *partial Umbell*, of an oblong, ovate fhape, running out to a point, oppofite, and often irregular.

CALYX glabra, perfiftens. *fig.* 1. *auct.*
COROLLA nulla.
NECTARIA quatuor, *corniculata*, fufca. *fig.* 2, 3.

CALYX fmooth, and permanent. *fig.* 1. *magnified.*
COROLLA wanting.
NECTARIES four, of a brownifh colour, each fur-nifhed with two little horns. *fig.* 2, 3.

STAMINA plerumque duo vifibilia ; ANTHERÆ didy-næ. *fig.* 4.
PISTILLUM : GERMEN fubrotundum, petiolatum, nu-tans ; STYLI tres ; STIGMA bifidum. *fig.* 5, 6.

STAMINA generally about two vifible ; ANTHERÆ double. *fig.* 4.
PISTILLUM : GERMEN roundifh, ftanding on a foot-ftalk, and drooping ; STYLES three ; STIGMA bifid. *fig.* 5, 6.

PERICARPIUM : CAPSULA tricocca, trilocularis.

SEED-VESSEL, a Capfule with three prominent fides, and three cavities.

SEMEN unicum in fingulo loculamento, nigrum, ru-gofum.

SEED : a fingle one in each cavity, black, and wrinkled.

This fmall and delicate fpecies of Spurge is often found in Corn-fields about *London*, efpecially on the *Surry* fide of the *Thames*, nor is it uncommon in many other parts of *England*.

It flowers in *July* and *Auguft.*

The *Tithymalus figetum longifolius* of RAY, confidered by Profeffor MARTYN in his *Plant. Cantab.* as the *fegetalis* of LINNÆUS, has by fome been thought to be no other than a large fpecimen of this plant ; Mr. HUDSON makes it a variety of the *platyphyllos*; fuppofing fuch a plant as the *fegetalis* to exift, it cannot be a variety of the *exigua* becaufe it has rough Capfules, and its leaves are by far too narrow for the *platyphyllos*, vid. JACQUIN *Fl. Auftr. V.* 3. & 4. who figures them both, and confiders them as diftinct fpecies.

Clematis Vitalba.

CLEMATIS VITALBA. TRAVELLER'S JOY.

CLEMATIS *Lin. Gen. Pl.* POLYANDRIA PENTAGYNIA.

Cal. o. *Petala* 4. rarius 5. *Sem.* caudata.

Raii Syn. Gen. 15. HERBÆ SEMINE NUDO POLYSPERMÆ.

CLEMATIS *Vitalba* foliis pinnatis : foliolis cordatis fcandentibus. *Lin. Syft. Veg. tab.* p. 426. *Spec. Pl.* 766.

CLEMATIS caule fcandente, foliis pinnatis, ovato lanceolatis, petalis coriaceis. *Haller Hd.* n. 1142.

CLEMATIS *Vitalba. Scopoli Fl. Carn.* p. 669.

CLEMATIS fylveftris latifolia. *Bauh. Pin.* 300.

CLEMATIS latifolia feu Atragene quibufdam. *J. B. II.* 125.

CLEMATIS fylveftris latifolia feu Viorna. *Parkinf.* 380.

VIORNA *Ger. emac.* 886. *Raii Syn.* 258. Great Wild Climber, or Traveller's Joy. *Hudfon Fl. Angl. ed.* 2. p. 238.

CAULES plurimi, perennantes, ope petiolorum, frutices arborefque vicinos, fcandentes; junioribus hirfutulis, flexilibus, purpureis, nodis incraffatis; per ætatem craffitie digiti feu pollicis, profunde fulcatis, exalbidis, fruticofis.

STALKS numerous, perennial, by means of the leaf-ftalks climbing the adjoining fhrubs and trees, the younger ones flightly hirfute, flexible, purple; the joints enlarged; by age attaining the thickneſs of the finger or thumb, deeply grooved, of a whitiſh colour and ſhrubby.

FOLIA pinnata, oppofita, patentia; pinnis duorum parium cum impari, remotis, cordatis, fubacuminatis, integris, ferratis, lobatifve, e viridi flavefcentibus, nitidulis, fubtus venolis.

LEAVES pinnated, oppofite, ſpreading, the pinnæ confifting of two pair with an odd one, heartſhaped, remote from each other, and running out to a point, either entire, ſerrated, or lobed, of a yellowiſh green colour, fomewhat ſhining; and veiny on the under fide.

PETIOLI contorti, vicem cirrhi fupplentes.

LEAF-STALKS twiſted, anſwering the purpoſe of a tendril.

RACEMI florales ex foliorum alis, conjugati, foliofi, ramofi, trifidi, dichotomi.

PLOWERING-BRANCHES proceeding from the alæ of the leaves, in pairs, leafy, branched, dividing firſt into three, and then into two ſmaller branches.

FLORES pallide fulphurei, odorati.

FLOWERS of a pale fulphur colour, and fweet-fcented.

CALYX nullus.

CALYX none.

COROLLA : PETALA quatuor, cruciata, oblonga, fubemarginata, patentia, fubrevoluta, villofa, fuperne pallide fulphurea, inferne virefcentia. *fig.* 1.

COROLLA: four PETALS crofs-ſhaped, oblong, flightly nicked at the end, ſpreading, fomewhat rolled back, villous, on the upper fide of a pale fulphur colour, underneath greeniſh. *fig.* 1.

STAMINA : FILAMENTA plurima, filiformia, fubcompreffa, alba, longitudine corollæ, erecta; ANTHERÆ oblongæ, albidæ. *fig.* 2.

STAMINA : FILAMENTS numerous, filiform, fomewhat flattened, white, the length of the corolla, upright; ANTHERÆ oblong, whitiſh. *fig.* 2.

PISTILLUM : GERMINA plurima, minima, fubrotunda, compreffa, definentia in tot STYLOS fubulatos, fericeos, longitudine ftaminum; STIGMATA fimplicia. *fig.* 3.

PISTILLUM : GERMINA numerous, very minute, roundiſh, flattened, terminating in as many tapering, filky STYLES, the length of the ftamina; STIGMATA fimple. *fig.* 3.

SEMINA plurima, nuda, fubrotunda, compreffa, caudata. *fig.* 4.

SEEDS numerous, naked, roundiſh, flatten'd, and terminated by a long feathered tail. *fig.* 4.

The *Clematis* [*] *Vitalba* [†] is a very common plant in the more fouthern parts of *Europe;* it delights in a fituation that is elevated, and in a foil that is chalky, hence it is found more plentifully in fome counties than in others; it is not frequent very near *London,* but abounds in the hedges around *Croydon,* and may be found, though more fparingly, about *Charlton,* alfo on the left hand fide of the road leading from New-Crofs Turnpike to *Lewisham,* near the fpot where the *Dipfacus pilofus* grows : it flowers in *Auguft,* and ornaments the hedges with its large branches of downy feeds; till the approach of winter.

Being a Climber, handfome both in its foliage and feeds, and rapid in its growth, it is often made ufe of for Arbours and Bowers in Gardens and Pleafure-Grounds; for this purpofe young plants fhould be chofen raifed from feeds. This quality, which is an ufeful one under proper reftrictions, often becomes a noxious one in hedges, where it is apt to fuffocate and deftroy thofe trees and fhrubs which are planted for defence.

School-boys often dry the ftems, when about the thicknefs of the finger, and draw fmoke through them inftead of cane. The Farmer alfo ufes the green ftalks to faften his gates with, &c.

HALLER quotes feveral authorities to fhew the *Clematis* poffeffed of confiderable acrimony, fufficient even to raife blifters on the fkin, for which purpofe it has fometimes been employed medicinally. A plant of the fame genus, viz. *Flammula Jovis (Clematis recta Linnei)* has been introduced into the laft edition of the *Edinburgh* Difpenfatory, on the authority of Dr. STORCK, who recommends the leaves to be externally applied, in fordid, ichorous, fungous, and cancerous ulcers, and caries of the bones; and preparations of them to be taken internally in the head-ach, nocturnal pains of the bones, venereal difeafe, itch and melancholy.

[*] So called from κλημα, *farmentum*, a vine twig.

[†] Quaſi *Vitis alba,* or white Vine.

[‡] Hence in fome parts of *England* the plant is called Old Man's beard.

RANUNCULUS REPENS. CREEPING CROWFOOT.

RANUNCULUS *Lin. Gen. Pl.* POLYANDRIA POLYGYNIA.

Cal. 5-phyllus. *Petala* 5 intra ungues poro mellifero. *Sem.* nuda.

Raii Syn. Gen. 15. HERBÆ SEMINE NUDO POLYSPERMÆ.

RANUNCULUS *repens* calycibus patulis, pedunculis fulcatis, ftolonibus repentibus, foliis compofitis. *Lin. Syft. Vegetab.* p. 430. *Fl. Suec.* n. 505. *Sp. Pl.* 779.

RANUNCULUS caule repente radicato, foliis femitrilobatis, lobis petiolatis. *Haller. Hift.* 1173.

RANUNCULUS repens. *Scopoli Fl. Carn.* n. 639.

RANUNCULUS pratenfis repens hirfutus. *Bauh. pin.* 179.

RANUNCULUS pratenfis repens. *Parkinf.* 329.

RANUNCULUS pratenfis etiamque hortenfis. *Ger. emac.* 951. *Raii Syn.* p. 247. Common creeping Crowfoot, or Butter-cups. *Hudfon Fl. Angl. ed.* 2. p. 240. *Lightfoot Fl. Scot.* p. 292.

RADIX plurimis fibris albentibus conftat.
CAULES ex una radice plerumque plures, pedales et ultra, variæ magnitudinis, pro ratione loci, teretiufculi, hirfuti, repentes.

PETIOLI longi, hirfuti, ad bafin dilatati.
FOLIA plerumque utrinque hirfuta (etiam glabra occurrunt) maculis albis fubinde notata, trilobata, lobis petiolatis, bi et tripartitis, lobulis acute dentatis.

RAMI floriferi erecti, fæpius biflori.

PEDUNCNLJ pubefcentes, ftriati.
CALYX: PERIANTHIUM pentaphyllum, foliolis ovatis, concavis, patentibus, pilofis, flavefcentibus, margine membranaceis, deciduis. *fig.* 1.
COROLLA: PETALA quinque, obcordata, patentia, flava, interne nitida. *fig.* 2.
NECTARIUM: *Squamula* parva, rotundata, ad bafin cujufvis petali. *fig.* 3.
STAMINA: FILAMENTA plurima, ultra triginta, receptaculo inferta, apice paululum dilatata; ANTHERÆ oblongo-lineares, compreffæ, incurvatæ, flavæ. *fig.* 4.
PISTILLUM: GERMINA plurima, in capitulum collecta, compreffa, erecta; STYLIS brevibus, acuminatis, apice reflexis terminati; STIGMATA fimplicia. *fig.* 5.
SEMEN compreffum, læve, mucronatum. *fig.* 6.

ROOT confifts of numerous whitifh fibres.
STALKS generally feveral from one root, a foot or more in length, varying in fize accord ng to the place of growth, roundifh, befet with rough hairs, and creeping.

LEAF-STALKS long, hairy and dilated at the bafe.
LEAVES generally hairy on both fides (fometimes they are found fmooth and fhining) frequently marked with white fpots, compofed of three lobes, or fmaller leaves which have footftalks, thefe are divided into two or three fegments, and fharply notched.
FLOWER-BRANCHES upright, generally fupporting two flowers.

FLOWER-STALKS downy and ftriated.
CALYX: a PERIANTHIUM of five leaves, which are ovate, concave, fpreading, hairy, yellowifh, membranous at the edge, and deciduous. *fig.* 1.
COROLLA: five PETALS, inverfely heart-fhaped, fpreading, yellow, fhining on the infide. *fig.* 2.
NECTARY a fmall roundifh *Scale* at the bafe of each petal. *fig.* 3.
STAMINA: FILAMENTS numerous, upwards of thirty, inferted into the receptacle, dilated a little at top; ANTHERÆ oblong and fomewhat linear, flattened, bent inward, and yellow. *fig.* 4.
PISTILLUM: GERMINA numerous, collected into a little head, flattened and upright; terminated by fhort, pointed STYLES, which turn back at top, STIGMATA fimple. *fig.* 5.
SEED flat, fmooth, with a fmall point. *fig.* 6.

The *Ranunculus bulbofus* is a very noxious plant in dry paftures, as the *acris* is in the moift, and fome of the beft meadows about town; but where the *repens* abounds, it is more mifchievous than either of thofe, and it is a plant fo general in its growth, that few meadows or paftures are entirely free from it; it differs from the other two Crowfoots, in having ftalks which run along the ground, and at every joint fending forth roots, and being a plant that will thrive in almoft any foil, it is very apt to become the principal plant of the pafturage, to the great detriment of the farmer, as cattle in general have the greateft averfion to the Crowfoots.

From the aftonifhing diverfity of foil and fituation in which this plant is found, the varieties which it affumes are almoft endlefs; by the Thames fide it will grow three or four feet high, with a ftem nearly as large as one's thumb; in barren, gravelly fields, it is entirely procumbent, with a ftalk not larger than a fmall wheat ftraw, but in all its various ftates I have ever found it to have a creeping ftalk, and this is a character which it does not lofe by cultivation. HALLER, milled by his pupil WILLICH*, who fince has retracted this error, fufpected it to be a variety of the *bulbofus*, but the *bulbofus* was never known to creep, this does wherever its ftalk can touch the ground.

Its principal time of flowering is in the month of June, but it may be found in bloffom during moft of the remaining fummer months.

Like the *acris* and *bulbofus* it is fometimes found double, but more rarely.

* XXXV *Ranunculum bulbofum* non in *repentem* mutari, ut in *Obfervationibus Botanicis* a. 1747. p. 4. fcripferam, fuperiores obfervationes docuerunt. Utraque planta diverfitates fuas conftantes retinet, ab I.c.c. LINNÆO nominibus fpecificis optime capeffita. *Obfervat. de plant. quibufd. Getting.* 1762.

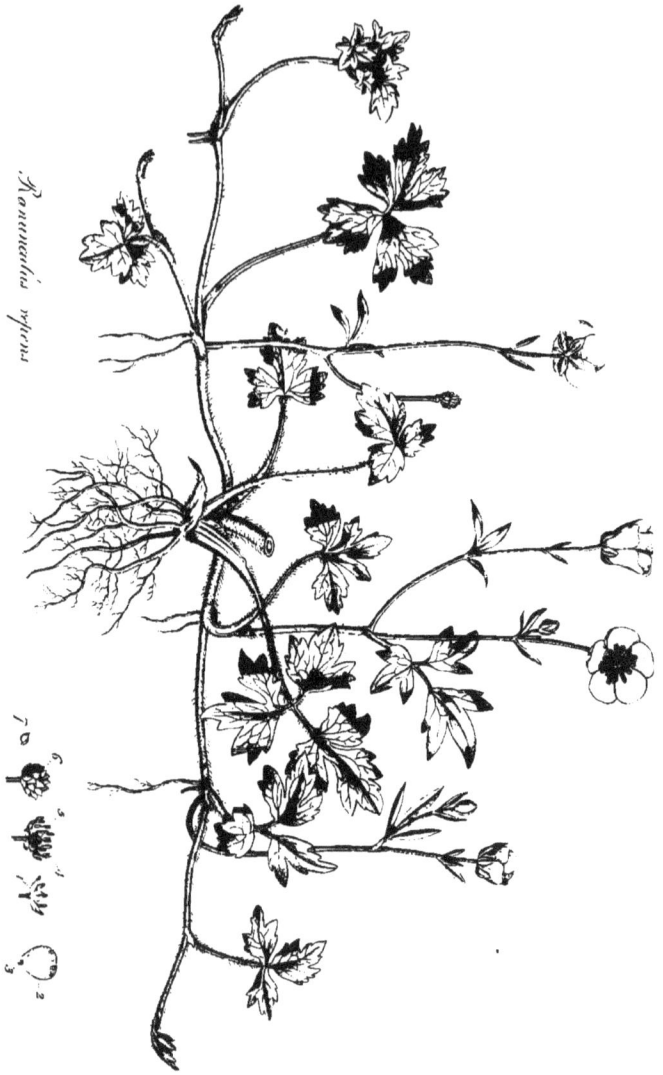

Ranunculus repens

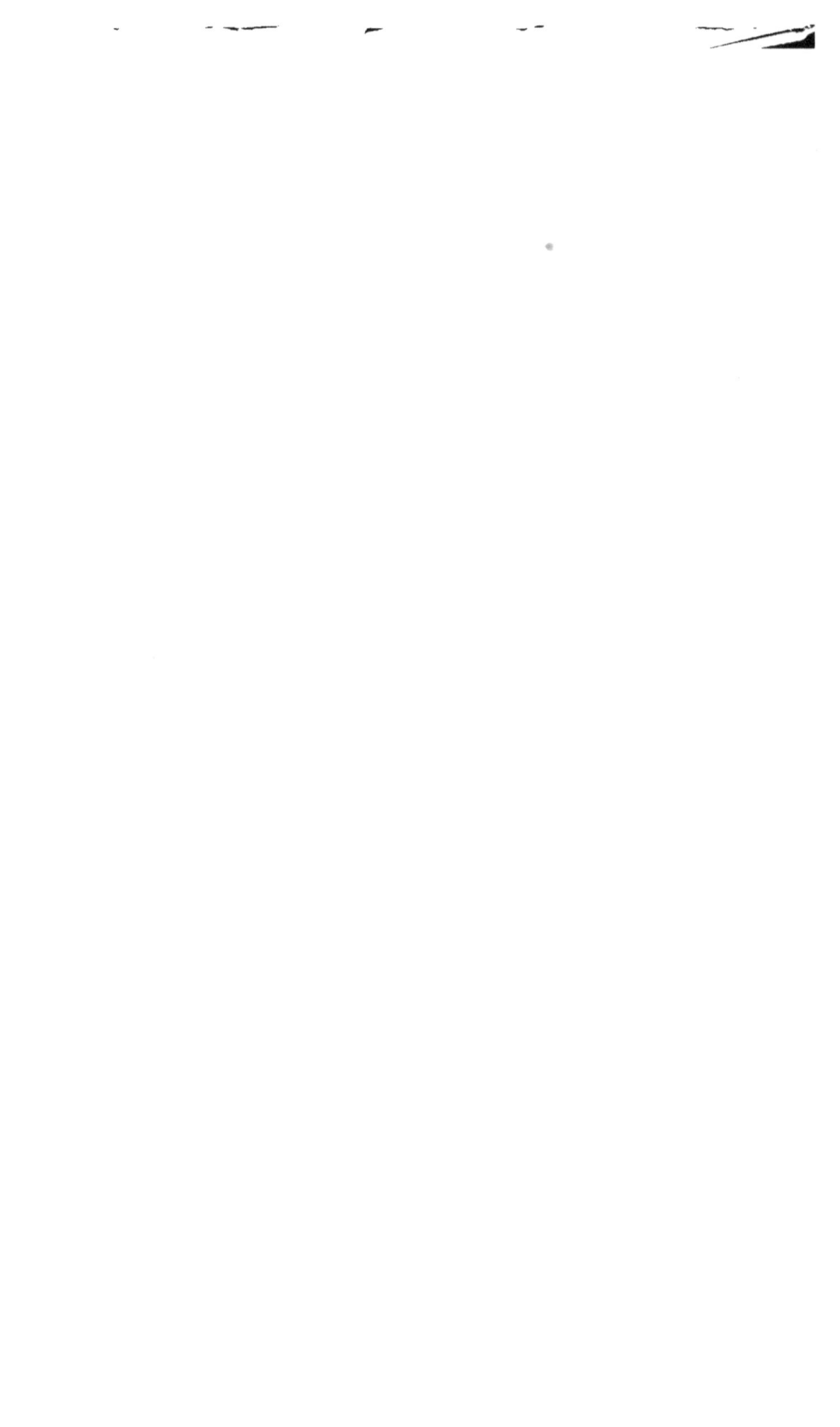

RANUNCULUS HEDERACEUS. IVY-LEAVED CROWFOOT.

RANUNCULUS *Lin. Gen. Pl.* POLYANDRIA POLYGYNIA.

Cal. 5-phyllus. *Petala* 5 Intra ungues poro mellifero. *Sem.* nuda.

Raii Syn. Gen. 15. HERBÆ SEMINE NUDO POLYSPERMÆ.

RANUNCULUS *hederaceus* foliis fubrotundis trilobis integerrimis, caule repente. *Lin. Syf. Vegetab.* p. 431. *Spec. Pl.* p. 781.

RANUNCULUS aquaticus hederaceus luteus. *Banb. Pin.* 180.

RANUNCULUS hederaceus rivulorum fe extendens, atra macula notatus. *I. B. III.* 782.

RANUNCULI aquatilis varietas altera. *Ger. emac.* 830.

RANUNCULUS hederaceus aquaticus. *Park.* 1216.

RANUNCULUS aquatilis hederaceus albus. *Raii Syn.* p. 249. *Hudfon Fl. Angl.* p. 243. *Lightfoot Fl. Scot.* p. 294. *Oeder Fl. Dan. ic.* 219.

RADICES plurimi, fimplices, albidi, in limum profunde demiffi.

ROOTS numerous, fimple, whitifh, penetrating deeply into the mud.

CAULES plurimi, teretes, folidi, geniculati, crafſiufculi, repentes, late diffuſi.

STALKS numerous, round, folid, jointed, thickifh, creeping, fpreading wide.

FOLIA petiolata, plerumque quinquelobata, fubcarnofa, nitida, avenia.

LEAVES ſtanding on foot-ſtalks, generally divided into five lobes, ſomewhat flefhy, fhining, and veinlefs.

PETIOLI ad bafin vagina magna, membranacea inftructi.

LEAF-STALKS at the bafe furnifhed with a large membranous fheath.

PEDUNCULI axillares, petiolis breviores, uniflori, primo erecti, peracta florefcentia verfus terram recurvati.

FLOWER-STALKS proceeding from the alæ of the leaves fhorter than the leaf-ſtalks, fupporting one bloffom, at firſt upright, after the flowering is over, turned back towards the ground.

FLORES parvi albi.

FLOWERS fmall and white.

CALYX: PERIANTHIUM pentaphyllum, foliolis ovatis, margine membranaceis, flavefcentibus. *fig.* 1.

CALYX: a PERIANTHIUM of five leaves, which are ovate, membranous and yellowifh at the edges. *fig.* 1.

COROLLA: petala quinque alba, oblonga, obtuſiufcula, calyce longiora. *fig.* 2.

COROLLA: five white PETALS of an oblong fhape, fomewhat obtufe, and larger than the calyx. *fig.* 2.

NECTARIUM: *Porus* nudus prope bafin cujufvis petali. *fig.* 3.

NECTARY: a naked pore at the bafe of each petal. *fig.* 3.

STAMINA: FILAMENTA quinque ad decem. ANTHERÆ oblongæ, flavæ. *fig.* 4.

STAMINA: FILAMENTS from five to ten. ANTHERÆ oblong and yellow. *fig.* 4.

SEMINA plurima, obtuſa, *fig.* 5. in capitula fubrotunda, vicine vulgaris magnitudine fere, denfe ftipata.

SEEDS numerous, blunt, *fig.* 5. crouded together in roundifh heads, about the fize of the common tare.

Neither LINNÆUS, HALLER, nor SCOPOLI, mention this plant in their refpective Flora's. OEDER figures it in the *Flora Danica*: the plant he gives us appears as if it had grown in water of fome depth; with us it is moſt commonly found fpreading widely on the mud of flow fhallow rivulets, efpecially where the foil is fandy.

It flowers during moſt of the fummer months.

We fometimes meet with the leaves having a dark fpot in the middle of each, and in fome fituations the bloffoms are much larger than in others.

Ranunculus hederaceus.

Galeobdolon Galeopsis

GALEOBDOLON GALEOPSIS. YELLOW ARCHANGEL.

GALEOBDOLON *Hudson Fl. Angl.* DIDYNAMIA GYMNOSPERMIA.

Raii Syn. Gen. 24. SUFFRUTICES ET HERBÆ VERTICILLATÆ.

GALEOBDOLON *luteum. Hudson. Fl. Angl. ed.* 2. p. 258.

GALEOPSIS *Galeobdolon* verticillis sexfloris; involucro tetraphyllo. *Lin. Syst. Vegetab.* p. 446. *Spec.* Pl. p. 810.

CARDIACA foliis petiolatis, cordatis, verticillis foliosis. *Haller. Hist.* 275.

LEONURUS *Galeobdolon. Scopoli Fl. Carn.* n. 705.

LAMIUM folio oblongo luteum. *Bauh. Pin.* 231. Lamium luteum. *Ger. emac.* 671. *Parkins* 605. *Raii Syn.* p. 240. Yellow Archangel or Dead Nettle.

GALEOPSIS *Galeobdolon. Lightfoot Fl. Scot.* p. 310.

RADIX perennis, inæqualis, fibras plurimas, majusculas, in terram demittens.

ROOT perennial, irregular, sending down several largish fibres.

CAULES plures, tetragoni, hirsutuli, *floriferi* suberecti, pedales seu bipedales, *steriles* peracta florescentia, in longum extenduntur, et postea humi repent.

STALKS several, four-cornered, somewhat hirsute: those producing flowers nearly upright, a foot or two feet high; those destitute of blossoms, after the flowering is over, are extended to a great length, and afterwards creep on the ground.

FOLIA opposita, petiolata, hirsutula, inæqualiter serrata, venosa, inferioribus cordatis, superioribus ovatis, acutis, sessilibus.

LEAVES opposite, standing on foot-stalks, slightly hirsute, unevenly serrated, and veiny; the lower ones heart-shaped, the upper ones ovate, pointed, and sessile.

FLORES verticillati, lutei.

FLOWERS growing in whirls, of a yellow colour.

VERTICILLI sex, decem aut duodecim flori.

WHIRLS containing from six to ten or twelve flowers.

CALYX: *Involucrum* verticillis subjectum, foliolis tot quot floribus, linearibus, acutis, rigidulis, ad lentem ciliatis, *fig.* 2.

CALYX: an *Involucrum* placed under the whirls, composed of as many leaves as there are flowers, the leaves linear, pointed, somewhat rigid, when magnified fringed at the edge. *fig.* 2.

CALYX: PERIANTHIUM monophyllum, subcampanulatum, quinquedentatum, hirsutulum, lineis decem elevatis notatum, alternis obsoletis, dentibus subæqualibus, acuminatis, superiore erecto, distanti, duobus inferioribus reflexopatulis. *fig.* 1.

CALYX: a PERIANTHIUM of one leaf, somewhat bell-shaped, having five teeth, slightly hirsute, marked with ten elevated lines, alternately faintest, the teeth nearly equal, having long points, the uppermost upright, and at a distance from the rest; the two lowermost spreading open and turned somewhat back. *fig.* 1.

COROLLA monopetala, ringens; *tubus* calyce paulo longior, intus purpureus et pilosus; *labium superius* erectum, longum, fornicatum, villosum, villisque ciliatum; *inferius* trifidum, laciniis inæqualibus, maculatis, media productiore. *fig.* 3, 4.

COROLLA monopetalous and ringent; *tube* a little longer than the calyx, purple and hairy within; *upper lip* upright, long, arched, villous, and edged with woolly hairs; the lowermost divided into three unequal segments which are spotted, the middle one longest. *fig.* 3, 4.

STAMINA: FILAMENTA quatuor, subulata, flava, sub labio superiore. ANTHERÆ bilobæ, purpurascentes. POLLEN albidum. *fig.* 5, 6.

STAMINA: four FILAMENTS tapering, of a yellow colour, under the upper lip. ANTHERÆ composed of two lobes and purplish. POLLEN whitish. *fig.* 5, 6.

PISTILLUM: GERMEN quadripartitum. STYLUS filiformis, purpureus. STIGMA bifidum, acutum. *fig.* 7.

PISTILLUM: GERMEN divided into four parts. STYLE filiform and purple. STIGMA bifid and pointed. *fig.* 7.

SEMINA quatuor, e fusco nigricantia, nitidula, triquetra, apice truncata in fundo calycis. *fig.* 9.

SEEDS four, of a brownish black colour, somewhat shining, three cornered, cut off at top, remaining in the bottom of the calyx. *fig.* 9.

LINNÆUS, though he enumerates this plant with the *Galeopsis* tribe, seems to think it not perfectly reconcileable with the rest. HALLER and SCOPOLI disagree in their opinions respecting it; the one considering it as a *Cardiaca*, the other as a *Leonurus*. Mr. HUDSON, whom we have followed in this instance, in the last edition of his *Flora Anglica*, makes a separate genus of it under the name of *Galeobdolon*; we adopt the trivial name of *Galeopsis* instead *luteum*, with a view of making as little innovation as possible in names.

It is always found in woods and shady places. In some parts of *England* it is frequent, but not in others: we find it tolerably plentiful in *Charlton*, and some other Woods about town, flowering in *May* and *June*.

The foliage is sometimes variegated, in which state I have observed it make a beautiful appearance in a garden. In cultivating this, as well as all other plants, attention should be paid to their natural place of growth.

Stachys arvensis.

STACHYS ARVENSIS. CORN STACHYS.

STACHYS *Lin. Gen. Pl.* DIDYNAMIA GYMNOSPERMIA.

Corollæ lab. super. fornicatum, lab. inferius lateribus reflexum: intermedia majore emarginata. *Stamina* deflorata versus latera reflexa.

Raii Syn. Gen. 14. SUFFRUTICES ET HERBÆ VERTICILLATÆ.

STACHYS *arvensis* verticillis sexfloris, foliis obtusis nudiusculis, corollis longitudine calycis, caule debili. *Lin. Syst. Vegetab.* p. 448. *Sp. Pl.* p. 814.

GLECOMA *arvensis. Lin. Fl. Suec. v.* 512. *Hudson Fl. Angl.* ed. 1. p. 224. upright Ground-ivy.

TRIXAGO foliis ovatis, cordatis, rotunde dentatis, hirsutis. *Haller. Hist.* n. 231.

SIDERITIS alsines trixaginis folio. *B. Pin.* 233.

SIDERITIS hederulæ folio. *Park.* 587.

SIDERITIS humilis lato obtuso folio. *Gerard. emac.* 699.

STACHYS arvensis minima. *Riv. Irr. Mon. icon. Raii Syn.* p. 242. *Hudson Fl. Angl.* ed. 2. p. 260. Corn Stachys. *Lightfoot Fl. Scot. p.* 314. petty Iron-wort or All-heal. *Oeder Fl. Dan. t.* 587.

RADIX annua, fibrosa.

ROOT annual and fibrous.

CAULIS spithamæus seu pedalis, inter segetes, ubi nobiscum sæpius invenitur, erectus (procumbens seu debilis rarius observatur) ramosus, tetragonus, hirsutus. *Rami* alterni, oppositi.

STALK a span or a foot in height, among the corn where it is generally found with us upright, (it is more rarely found weak or procumbent), branched, four cornered, hirsute. *Branches* alternately opposite.

FOLIA opposita, petiolata, ovato-cordata, obtusa, venosa, hirsuta, obtuse serrata, summa sessilia.

LEAVES opposite, standing on foot-stalks, ovate, with an heart-shaped base, obtuse, veiny, hirsute, bluntly serrated, the uppermost ones sessile.

FLORES parvi, carnei, verticillati, spicati.

FLOWERS small, flesh-coloured, growing in whirls, and forming a spike.

VERTICILLI circiter sexflori.

WHIRLS containing six flowers in each.

CALYX: PERIANTHIUM quinque-dentatum, persistens, patens, hirsutum. *fig.* 1.

CALYX: PERIANTHIUM with five teeth, permanent, projecting horizontally, and hairy. *fig.* 1.

COROLLA ringens, parva, calyce paulo longior, pallide purpurea, labio superiore breviore, obtuso, integro, inferiore trifido, laciniis lateralibus brevioribus, media majori, rotundata. *fig.* 2.

COROLLA ringent, small, a little longer than the calyx, of a pale purple colour, the upper lip short, blunt, and entire; the lower one divided into three segments, of which the two side ones are the shortest, the middle one large and roundish. *fig.* 2.

STAMINA: FILAMENTA quatuor, quorum duo breviora, inferne alba, superne purpurea; ANTHERÆ nigricantes; POLLEN flavum. *fig.* 3.

STAMINA: four FILAMENTS, two of which are shorter than the rest, white below, and purple above; ANTHERÆ blackish; POLLEN yellow. *fig.* 3.

PISTILLUM: GERMEN quadripartitum, viride. STYLUS longitudine staminum. STIGMA bifidum, acutum. *fig.* 4.

PISTILLUM: GERMEN divided into four parts, of a green colour. STYLE the length of the stamina. STIGMA bifid and pointed. *fig.* 4.

SEMINA quatuor intra calycem, virescentia, nigro punctata. *fig.* 5.

SEEDS four within the calyx, of a greenish colour, dotted with black. *fig.* 5.

Botanists, both ancient and modern, seem to have been at a loss with what genus of plants they should class this small inhabitant of our Corn-fields. At various times, and by various authors, it has been called a *Sideritis*, a *Lamium*, a *Marrubiastrum*, a *Glechoma*, a *Stachys*, and a *Trixago*. It appears neither to possess the striking characters of any other genus, nor to afford sufficient distinction to form a genus by itself. As a species, however, there is no difficulty about it; its place of growth, the smallness of its flowers, and form of its leaves, obviously distinguish it from any other English plant.

It is not uncommon in the Corn-fields about the *Oak of Honour* and *Coomb* Woods, and elsewhere about London; it flowers in *June, July,* and *August.*

Prunella vulgaris

PRUNELLA VULGARIS.　　SELF-HEAL.

PRUNELLA *Lin. Gen. Pl.* DIDYNAMIA GYMNOSPERMIA.

Filamenta bifurca: altero apice antherifero. *Stigma* bifidum.

Raii Syn. Gen. 14. SUFFRUTICES ET HERBÆ VERTICILLATÆ.

PRUNELLA *vulgaris* foliis omnibus ovato-oblongis ferratis petiolatis. *Lin. Syst. Vegetab.* p. 458. *Sp. Pl.* 837. *Fl. Suec.* 540.

BRUNELLA foliis ovatis oblongis calycibus fuperne truncatis. *Haller. Hift.* n. 277.

BRUNELLA *vulgaris. Scopoli Fl. Carn.* 715.

PRUNELLA major folio non diffecto. *Bauh. pin.* 260.

PRUNELLA *Gerard. emac.* 631.

PRUNELLA vulgaris. *Parkinf.* 1680. *Raii Syn.* p. 238.　Common Self-heal.　*Hudfon. Fl. Angl.* ed. 2. p. 264.　*Lightfoot Flor. Scot.* p. 321.

RADIX annua, fibrofiffima.

CAULIS dodrantalis, pedalis et ultra, erectus, ad bafin ufque ramofus, geniculatus, geniculis inferioribus radicantibus, fubquadratus, utrinque canaliculatus, hirfutus, purpurafcens.

RAMI cauli fimiles, adfcendentes, alterne oppofiti, inferioribus longioribus.

FOLIA oppofita, petiolata, ovata, plana, patentia, punctis prominulis fcabriufcula, obfolete dentata potius quam ferrata.

PETIOLI fuperne canaliculati, marginibus hirfutis.

SPICÆ terminales, feffiles, fubcylindricæ, pollicares, truncatæ, e floribus verticillatis breviter pedicellatis et bracteis conftantes.

BRACTEÆ cordatæ, latæ, acuminatæ, fubdiaphanæ, venofæ, ciliatæ, fubtus hirfutæ, fupra glabræ.

VERTICILLI fexflori.

CALYX: PERIANTHIUM tubulatum, bracteis paulo longius, biangulatum, fupra glabrum, trinerveum, fubtus ftriatum, angulis ciliatis, bilabiatum; labio fuperiore depreffo-plano, truncato, crenato, minutim tridentato, lateribus planis, inferiore anguftiore, bifido, laciniis acuminatis. *fig.* 1.

COROLLA violacea, caduca, monopetala, ringens, pars tubulofa calyce paulo longior, labium fuperius concavum, integrum, fuperne villofum, inferius trifidum, laciniis lateralibus integris, deflexis, intermedia erecta, rotundata, denticulata. *fig.* 2.

STAMINA: FILAMENTA quatuor, fubulata, violacea, bifurca, quorum duo breviora; ANTHERÆ bilobæ, lobis divergentibus, furcâ interiore infidentes. *fig.* 3.

PISTILLUM: GERMEN quadripartitum, glandulâ crenatâ cinctum; STYLUS filiformis, longitudine ftaminum; STIGMA bifidum, acutum. *fig.* 4, 5, 6, 7.

SEMINA quatuor, ovata, parva, obtufe angulata, glabra, fufca, mucrone albo terminata. *fig.* 8.

‡ ROOT annual, and exceedingly fibrous.

STALK from half a foot, to a foot or more in height, upright, branched quite down to the bottom, jointed, the lowermoft joints fending down roots, fomewhat fquare, with a deep groove on each fide, rough and purplifh.

BRANCHES like the ftalk, bending upward, alternately oppofite, the lower ones longeft.

LEAVES oppofite, ftanding on footftalks, ovate, flat, fpreading, rough with little prominent points, faintly indented on the edge, father than fawed.

LEAF-STALKS hollow above, and hairy on the edges.

SPIKES terminal, feffile, fomewhat cylindrical, about an inch in length, as if cut off at top, compofed of floral leaves and flowers ftanding on fhort footftalks.

FLORAL-LEAVES heartfhaped, broad and pointed, fomewhat tranfparent, veiny, edged with hairs, hirfute on the under fide, fmooth and fhining above.

WHIRLS fix flowers in each.

CALYX: a tubular PERIANTHIUM, fomewhat longer than the floral-leaves, angular on each fide, fmooth above, with three faint ribs, ftriated below, the angles edged with hairs, compofed of two lips, the uppermoft of which is flat, and fomewhat depreffed, cut off at top, terminated by three fmall teeth, the fides flat, the lower lip narrower, bifid, the fegments long and pointed. *fig.* 1.

COROLLA of a violet colour, eafily falling, monopetalous and ringent, the tubular part a little longer than the calyx, the upper lip hollow, entire, and villous above, the lower lip divided into three fegments, of which the fide-ones are entire and bend downward, the middle one upright, round, and finely toothed. *fig.* 2.

STAMINA: four FILAMENTS, tapering, of a violet colour, forked at top, of which two are longer than the others; ANTHERÆ compofed of two lobes, which diverge and fit on the inner fork. *fig.* 3.

PISTILLUM: GERMEN divided into four parts, furrounded by a notched gland; STYLE filiform, the length of the ftamina; STIGMA bifid and pointed. *fig.* 4, 5, 6, 7.

SEEDS four, ovate, fmall, obtufely angular, fmooth, brown, and terminated by a white point. *fig.* 8.

In many of the natural claffes of plants, the feveral *genera* approximate fo much, that it is difficult to find out a character which fhall obvioufly diftinguifh them; fuch a character is however afforded in the prefent plant. If the filaments are examined, each of them will be found forked at the extremity, and the anthera fuftained on the innermoft divifion of the fork; befides this curious and uncommon ftructure, the middle fegment of the lower lip is finely toothed. This is noticed by TOURNEFORT in his figures of this genus, but the former wholly omitted. It did not however efcape the penetrating LINNÆUS, who confiders it as the effential character of the *Prunella*; nor is the calyx in this plant undeferving of our attention, whether we confider its ftructure, or the fingular manner in which it clofes up to preferve the feeds.

The *Self-heal* has an herbaceous roughifh tafte, and hence ftands recommended in alvine fluxes; it has been principally celebrated as a vulnerary, whence its name, and in gargarifms for apthæ, and inflammations of the fauces. *Lewis's Difp.* p. 205.

It grows very commonly in meadows and paftures, and flowers in June and July. Its bloffoms, which are ufually of a violet colour, are fometimes found white, and fometimes red.

According to LINNÆUS's experiments, kine, goats, and fheep, eat it; but horfes refufe it.

SCUTELLARIA MINOR. SMALL HOODED-WILLOW HERB.

SCUTELLARIA *Lin. Gen. Pl.* DIDYNAMIA GYMNOSPERMIA.

Calyx ore integro : poſt floreſcentiam clauſo, operculato.

Raii Syn. Gen. 14. SUFFRUTICES ET HERBÆ VERTICILLATÆ.

SCUTELLARIA *minor* foliis cordato-ovatis ſubintegerrimis, floribus axillaribus. *Lin. Syſt. Vegetab.* p. 457. *Sp. Plant.* p. 835.

CASSIDA paluſtris minima flore purpuraſcente. *Tourn. Hiſt.* p. 410.

LYSIMACHIA galericulata minor. *Raii Hiſt.* 572. *Syn.* 244. The leſſer-hooded Looſeſtriſe.

GRATIOLA latifolia. *Gerard. emac.* 585.

GRATIOLA latifolia ſen noſtras minor *Parkins.* 221. *Hudſon. Fl. Angl. ed.* 2. p. 265. *Lightfoot Fl. Scot.* p. 320.

RADIX perennis, repens, alba.

CAULIS erectus, palmaris, ſeſquipalmaris, et ultra, ſimplex, ſeu ramoſus, in horto ramoſiſſimus evadit, tetragonus, rigidulus, ſubhirſutus, baſi purpuraſcens.

FOLIA oppoſita, ſubſecunda, breviſſime petiolata, cordato-ovata, obtuſa, ſubintegerrima, in opacis glabra, in apticis hirſuta, et ſubinde rubentia.

FLORES parvi, carnei, axillares, bini, pedunculati, pedunculis calyce longioribus.

CALYX : PERIANTHIUM monophyllum, breviſſimum, hirſutum, ore bilabiato, integro, ſquamula incumbente operculi inſtar clauſuli. *fig.* 1.

COROLLA monopetala, ringens, labium ſuperius trifidum, ſuperne villoſum, laciniis ſubæqualibus, intermedia concavâ emarginatâ, lateralibus remotiuſculis, nec intermediæ ſubjectis ſicut in galericulata. *fig.* 2. inferius latius, emarginatum, punctis et lineolis rubris pulchre variegatum. *fig.* 3.

STAMINA : FILAMENTA quatuor, alba, ſub labio ſuperiore recondita, quorum duo longiora. ANTHERÆ minimæ, albæ, ad lentem hirſutulæ. *fig.* 4.

PISTILLUM : GERMEN quadripartitum. STYLUS albus, longitudine ſtaminum, ſuperne paululum incraſſatus. STIGMA ſimplex, incurvatum, acuminatum. *fig.* 5.

PERICARPIUM nullum, calyx continens

SEMINA quatuor, ſubrotunda, ad lentem aſpera. *fig.* 6.

NECTARIUM glandula aurantiaca ad baſin germinis. *fig.* 7.

ROOT perennial, creeping, and white.

STALK upright, a hand's breadth, ſix inches, or more in height, ſimple or branched, in the garden becoming very much ſo, four-cornered, ſomewhat rigid and hirſute, purpliſh at the baſe.

LEAVES oppoſite, with a tendency to grow one way, ſtanding on very ſhort foot-ſtalks, heart-ſhaped, ovate, obtuſe, almoſt entire on the edges, in ſhady ſituations ſmooth, in expoſed ones hirſute, and ſometimes reddiſh.

FLOWERS ſmall, of a pale red colour, growing by pairs in the alæ of the leaves, ſtanding on foot-ſtalks longer than the calyx.

CALYX : a PERIANTHIUM of one leaf, very ſhort, hirſute, the mouth compoſed of two lips and entire, with a ſmall ſcale like a lid reſting on it. *fig.* 1.

COROLLA monopetalous, and ringent, the upper lip trifid, and villous above, the ſegments nearly equal, the intermediate one hollow and nicked, the ſide ones ſomewhat diſtant from each other, not placed under the intermediate one as in the galericulata. *fig.* 2. the lower lip broader, with a ſingle notch, and prettily variegated with red lines and dots. *fig.* 3.

STAMINA : four FILAMENTS, of a white colour, hid under the upper lip, two of which are longer than the other two. ANTHERÆ very minute, white, and ſlightly hirſute, when magnified. *fig.* 4.

PISTILLUM : GERMEN divided into four parts. STYLE white, the length of the ſtamina, thickened a little above. STIGMA ſimple, a little hooked, and pointed. *fig.* 5.

SEED-VESSEL none, the calyx containing four ſeeds.

SEEDS of a roundiſh ſhape, appearing rough when magnified. *fig.* 6.

NECTARY : an orange-coloured gland at the baſe of the germen. *fig.* 7.

GERARD, our countryman, appears to have been the diſcoverer of this ſpecies, to which he gives the names of *Gratiola latifolia*. PARKINSON, who conſidered it as a plant peculiar to this country, adds to them the epithet of *noſtras*. TOURNEFORT, afterwards enumerating the plants growing about *Paris*, mentions it as growing with them in ſeveral places ; neverthelefs it is not common throughout Europe. Indeed LINNÆUS, in his *Species Plantarum*, expreſsly ſays, *habitat in Anglia*. GERARD's referring a plant to another genus, ſo obviouſly ſimilar in its parts of fructification to what was then called *Lyſimachia galericulata* ſhews, that little attention was paid to thoſe parts at that time ; nor was he very fortunate in applying to the leaves of ſo ſmall a plant the name of *latifolia*.

Had LINNÆUS frequently ſeen and cultivated this plant, as we have had opportunity of doing, he would have expreſſed no doubt of its being too nearly related to the *galericulata* and *baſtifolia*. *Vid. Spec. Plant.*

In the time of old GERARD, this ſpecies was found on *Hampſtead Heath*, and on ſome of the boggy parts of that Heath it ſtill exiſts ; but is more plentifully met with in ſimilar ſituations, on *Putney* and *Shirley* Commons.

It flowers in July, Auguſt, and September. On *Shirley* Common we have ſeen it much larger than the ſpecimen figured ; and this ſummer found a ſingle plant which had blue flowers.

Scutellaria minor

Orobanche major.

OROBANCHE *Lin. Gen. Pl.* DIDYNAMIA ANGIOSPERMIA.

Cal. bifidus. *Cor.* ringens. *Caps.* unilocularis, bivalvis, polyfperma. *Glandula* fub bafi germinis.

Raii Syn. Gen. 18. HERBÆ FRUCTU SICCO SINGULARI FLORE MONOPETALO.

OROBANCHE *major* caule fimpliciffimo pubefcente, ftaminibus fubexfertis. *Lin. Syft. Vegetab.* p. 497. *Spec. Pl.* p. 882. *Fl. Suec.* n. 561.

OROBANCHE caule fimplici, ftipula unica, calyce quadrifida. *Haller. Hist.* 295.

OROBANCHE *major. Scopoli Fl. Carn.* n. 782.

OROBANCHE major Garyophyllum olens. *Bauh. pin.* 87.

RAPUM Geniftæ *Ger. emac.* 1311. *Parkinf.* 229. *Raii Syn.* p. 288. Broom-rape. *Hudfon. Fl. Angl.* ed. 2. p. 266. *Lightfoot Flor. Scot.* p. 332.

RADIX Spartii fcoparii radicibus plerumque adnafcitur.

ROOT generally grows to the roots of the common Broom.

CAULIS pedalis ad fefquipedalem, erectus, fimplex, fiftulofus, teretiufculus, fulcato-ftriatus, villofus, purpurafcens, fquamis fparfis, marcidis, tectus, ad bafin bulbofus, bulbo fquamofo.

STALK from a foot to a foot and a half in height, upright, fimple, hollow, roundifh, fomewhat channeled, villous, of a purplifh colour, covered with withered fquamæ, bulbous at bottom, the bulb fquamous.

FLORES fpicati, feffiles, purpurafcentes plerumque, aliquando etiam flavefcentes, fpica primo acuta, dein fubcylindrica.

FLOWERS growing in fpikes, feffile, for the moft part purplifh, fometimes alfo yellowifh, the fpike is at firft pointed, and then becomes cylindrical.

CALYX : PERIANTHIUM monophyllum, erectum, quadrifidum, poftice ad bafin ufque divifum, et antice magis profunde quam ad latera, villofum, albefcens, laciniis feu dentibus fubæqualibus, ferrugineis. *fig.* 1.

CALYX : a PERIANTHIUM of one leaf, upright, divided into four fegments, pofteriorly quite down to the bafe, and anteriorly more deeply that at the fides; villous, whitifh, the fegments or teeth nearly equal, and ferruginous. *fig.* 1.

COROLLA : monopetala, ringens, plerumque purpurafcens, ad ferrugineum colorem accedens, perfiftens; *tubus* inclinatus, amplus, ventricofus; *faux* hians; *labium fuperius* concavum, obfolete trifidum, fubcarinatum, externe vifcidum; *labium inferius* trifidum, lacinulâ mediâ productiore, concava. *fig.* 2.

COROLLA monopetalous, ringent, generally purplifh, approaching to the colour of rufty iron; *tube* bending downward, large, bellying out; *mouth* gaping; *upper-lip* hollow, obfoletely trifid, fomewhat keeled, and externally vifcid; the lower lip trifid and hollow, the middle fegment the longeft. *fig.* 2.

STAMINA : FILAMENTA quatuor, fubulata, alba, fub labio fuperiore recondita, quorum duo longiora; ANTHERÆ leviter cohærentes, ftramineæ, didymæ, ovatæ, mucronibus terminatæ. *fig.* 3.

STAMINA : four FILAMENTS, tapering, white, hid under the upper lip, two of which are longer than the reft; ANTHERÆ flightly cohering, of a ftraw colour, double, ovate, each cavity terminating in a point. *fig.* 3.

PISTILLUM : GERMEN oblongum, bafi antice tribus glandulis, protuberantibus, flavis, melleum liquorem copiofe fundentibus, notatum; STYLUS pubefcens, ftaminibus paulo longior, purpurafcens, apice inflexus; STIGMA femibifidum, obtufum, craffiufculum, flavum. *fig.* 4.

PISTILLUM : GERMEN oblong, marked anteriorly at the bafe with three protuberant yellow glands, which pour forth plentifully a fweet liquor; STYLE downy, a little longer than the flamina, purplifh, bent down at top; STIGMA half-divided, obtufe, thickifh, and yellow. *fig.* 4.

PERICARPIUM : CAPSULA ovato-oblonga, acuminata, unilocularis, bivalvis. *fig.* 5.

SEED-VESSEL an ovate, oblong *capfule*, terminating in a point, of one cavity and two valves. *fig.* 5.

SEMINA numerofa, minima; *receptacula* quatuor, linearia, lateralia, adnata. *fig.* 6, 7.

SEEDS numerous and very fmall; *receptacles* four, linear, growing to the fides of the capfule. *fig.* 6, 7.

The literal Englifh tranflation of the Greek word *Orobanche* is *Strangle-tare* * : this term has been given by DIOSCORIDES to one plant, and by THEOPHRASTUS to another; ours is undoubtedly the *Orobanche* of DIOSCORIDES, (as it accords with his defcription †), and alfo of PLINY. The *Orobanche* of THEOPHRASTUS muft have been fome

---

* ΟΡΟΒΑΓΧΗ, *Ervangina Gonos, quid* ὄρβον ἄγχει, *ervum ftrangulat, dicitur.*

† " Orobanche caulicolus eft fefquipedalis, et interdum major, fubruber, hirfutus, tener, fine folio, pinguis: flore fuhalbido, aut in luteum " vergente : radice digiti craffitudine, et eum ariditate flacceftcit caulis, fiftulofa. Hæc inter quædam legumina nafci conftat, et ea ftrangulare, " unde Orobanche fibi cognomentum ufurpavit. Eftur, ut olus, cruda, et in patiniis, afparagi modo, decocta. Legumentis addita, conieoctio-" nem accelerare creditur." *Matth. in Diofc.*

climbing plant, as is evident from his words, which are thus quoted by MATTHIOLUS, " Ervum necat amplexu " complexuque suo," whereas the *Orobanche* of DIOSCORIDES, according to the same author, by its presence only, " suâ tantum præsentiâ," destroys the Legumina, Corn, Hemp, and Flax which grow near it, and from which property of choaking and devouring the neighbouring plants, MATTHIOLUS says they called it in some parts of Italy the *Wolf plant*; its pernicious effects in this respect are confirmed by a later Italian writer MICHELI, who mentions its being proscribed in Tuscany by public edict.

The most remarkable circumstance in the œconomy of this plant, is its growing from the roots of others; CASPAR BAUHINE afserts, that it is always attached to the fibres *(fibris)* of some plant near it; several of my botanic friends are of opinion that it is not parasitical in all situations, especially in corn-fields, but they have not confirmed their opinions by actual and repeated examinations, which are still wanting. The plants from the roots of which CASP. BAUHINE observed it to grow were the *Spartium Scoparium,* common Broom, *Genista tinctoria,* Woodwaxen, *Hieracium subaudium,* shrubby Hawkweed, *Trifolium,* Trefoil, (no particular species is mentioned), and *Orobus tuberosus,* Wood Pea; all of these (one excepted) are leguminous plants, to which it is observed to have the strongest attachment; I have met with it but rarely about London, excepting one spot, in which it might be said truly to abound; this was a small, hilly, barren field, covered partly with furze and broom, on the left-hand side of the road, within about two miles of Kingston, Surry, about half a mile beyond the Robin Hood and Turnpike, in which field the Botanist will find several other rare plants. I have also seen it on Hampstead Heath, and on the borders of Charlton Wood; in these several situations it grew either out of the roots of Broom or Furze, but chiefly the former. My late gardener ROBERT SQUIUS once brought me out of Surry a very strong plant of *Orobanche,* which had grown in a corn field on the root of the *Centaurea Scabiosa:* I planted both roots in my garden, in the situation they were found; the *Knapweed* grew, but the *Broom-rape* died. Several persons have informed me of their having found it to be parasitical on the roots of *Clover,* in particular Dr. GOODENOUGH and Mr. RUGGLES of Cobham; Mr. THOMAS WHITE once related to me his having observed a small Orobanche growing on walls, &c. in Pembrokeshire, and that the decayed floor of an old castle in particular was almost covered with it; he also noticed, that in some of the western counties this plant was so common as to give the mowers cause of complaint.

The *Orobanche* appears to vary in size according to the size of the root it grows on, the stem being sometimes almost as thick as one's thumb, and at other times not much larger than a wheat-straw; the flowers vary much in their colour, but are mostly dead-purple or yellowish.

The seed of this plant is remarkably small, hence it will be extremely difficult to shew its vegetation by experiment, more especially as it requires a very particular soil and situation, yet no one can doubt but the plant is propagated thereby; it must first vegetate on the earth, then the radicle, which is sent downward, finding a proper root, attaches itself to it, quits its parent earth, and becomes parasitical.

It has a faint smell of cloves, and is said to be a strong astringent and useful vulnerary.

LINNÆUS remarks that *Sweden* is too cold for it to flourish with them.

Antirrhinum Orontium.

ANTIRRHINUM *Lin. Gen. Pl.* DIDYNAMIA ANGIOSPERMIA.

> *Cal.* 5-phyllus. *Corollæ* bafis deorfum prominens, nectarifera. *Capfula* 2-locularis.

Raii Syn. Gen. 18. HERBÆ FRUCTU SICCO SINGULARI FLORE MONOPETALO.

ANTIRRHINUM *Orontium* corollis ecaudatis, floribus fubfpicatis, calycibus corolla longioribus. *Lin. Syft. Vegetab.* p. 466. *Sp. Pl.* p. 860. *Fl. Suec.* n. 559.

ANTIRRHINUM foliis ellipticis obtufis, floribus fparfis, calycibus fubulatis longiffimis, calcare bréviffima. *Haller. Hift.* n. 334.

ANTIRRHINUM *Orontium. Scopoli Fl. Carn.* n. 774.

ANTIRRHINUM anguftifolium fylveftre. *Bauh. Hift.* III. 464.

ANTIRRHINUM arvenfe majus. *Bauh. pin.* 212.

ANTIRRHINUM fylveftre medium. *Parkinf.* 1334. *Raii Syn.* p. *283. The leffer wild Calf's Snout or Snapdragon. *Hudfon. Fl. Angl. ed.* 2. p. 274.

RADIX annua, fimplex, rigida, albida, fibrofa, fibris paucis, patentibus.

ROOT annual, fimple, ftiff, whitifh, fibrous, fibres few and fpreading.

CAULIS fpithamæus, aut pedalis, fimplex feu ramofus, erectus, teres, pilofus, pilis patentibus, fubvifcidis.

STALK from a fpan to a foot in height, fimple or branched, upright, round, hairy, the hairs fpreading and fomewhat vifcid.

FOLIA ima oppofita, fubconnata, fuperiora alterna, lineari-lanceolata, carinata, deflexa, integerrima, hirfutula.

LEAVES of a fhape betwixt linear and lanceolate, keeled, turning downward, entire at the edge, flightly hairy, the lowermoft oppofite, flightly uniting at the bafe, the uppermoft alternate.

FLORES pauci, axillares, feffiles, rubelli, vix fpicati, nifi in fpeciminibus majoribus.

FLOWERS few, growing from the alæ of the leaves, feffile, of a bright red colour, fcarcely forming a fpike, unlefs in large fpecimens.

CALYX: PERIANTHIUM quinque-partitum, perfiftens, laciniis linearibus, carinatis, hirfutis, longitudine corollæ, fuperiore longiore.

CALYX: a PERIANTHIUM deeply divided into five fegments, and permanent; the fegments linear, keeled, hairy, the length of the corolla, the uppermoft fegment longeft.

COROLLA monopetala, rofea, venis faturatioribus ftriata, inferne et fuperne pubefcens; *labium* fuperius bifidum, lateribus reflexum, inferius trifidum, laciniis deflexis, intermedia minore, *Nectarium* breviffimum, obtufum. *fig.* 1.

COROLLA monopetalous, rofe coloured, ftriped with veins of the fame colour but deeper, both above and below flightly hairy, the upper lip bifid, the fides turning back, the lower lip trifd, the fegments turning down, the middle one the fmalleft; *Nectary* very fhort and blunt. *fig.* 1.

STAMINA: FILAMENTA quatuor, filiformia, purpurafcentia, glabra, quorum duo breviora; ANTHERÆ flavæ, bilobæ, conniventes. *fig.* 2.

STAMINA: four FILAMENTS, thread-fhaped, purplifh, fmooth, two of which are fhorter than the others; ANTHERÆ yellow, clofing together, compofed of two lobes. *fig.* 2.

PISTILLUM: GERMEN fubrotundum, villofum; STYLUS fubulatus, villofus, apice paululum inclinatus; STIGMA obtufum, flavum. *fig.* 3.

PISTILLUM: GERMEN roundifh, and villous; STYLE tapering, villous, bending a little downward at top; STIGMA blunt and yellow. *fig.* 3.

PERICARPIUM: CAPSULA pedunculata, erecta, hirfuta, apice triforaminofa, bilocularis. *fig.* 4.

SEED-VESSEL: a CAPSULE ftanding on a footftalk, upright, hairy, having two cavities and three apertures at top. *fig.* 4.

SEMINA plurima, nigricantia, angulata. *fig.* 5.

SEEDS numerous, blackifh and angular, *fig.* 5.

This fpecies of *Antirrhinum* is diftinguifhed from all the others, the *majus* excepted, by having no fpur to the corolla, and from the *majus* by being an annual, and having long, pointed leaves to the calyx, which in that plant are fhort and obtufe.

It grows in tolerable plenty in Batterfea Corn-fields, where it flowers in July and Auguft.

The feed-veffel, when fully ripe, is a curious reprefentation of the fkull of a Quadruped.

Raphanus Raphanistrum.

RAPHANUS RAPHANISTRUM. WILD RADISH.

RAPHANUS *Lin. Gen. Pl.* TETRADYNAMIA SILIQUOSA.

> *Cal.* claufus. *Siliqua* torofa, fubarticulata, teres. *Glandulæ* melliferæ 2 inter ftamina breviora et piftillum, totidem inter ftamina longiora et calycem.

Raii Syn. Gen. 21. HERBÆ TETRAPETALÆ SILIQUOSÆ ET SILICULOSÆ.

RAPHANUS *Raphaniftrum* filiquis teretibus articulatis lævibus unilocularibus. *Lin. Syft. Veget.* p 504. *Sp. Pl.* 935. *Fl. Suec.* n. 612. *Haller. Hift.* n. 468.

RAPHANISTRUM fegetum, flore luteo vel pallido. *Tournef. Inft.* 240.

RAPISTRUM flore luteo, filiqua glabra articulata. *Raii Syn.* p. 296.

RAPHANUS fylveftris. *Ger. emac.* 240. *Hudfon Fl. Angl.* p. 289. *Lightfoot Fl. Scot.* p. 362.

RADIX annua, fimplex, fibrofa, pallide fufca.

ROOT annual, fimple, fibrous, and of a pale brown colour.

CAULIS pedalis ad fefquipedalem, erectus, teres, folidus, hifpidus, glaucus, ad bafin ufque plerumque ramofus, ramis fæpe longitudine caulis, ad bafin purpureis.

STALK from a foot to a foot and a half in height, upright, round, folid, hifpid, glaucous, generally branched quite down to the bottom, branches often as long as the ftalk, and purple at bottom.

FOLIA petiolata, pinnatifida, alterna, fcabra, inferioribus pinnarum quatuor aut quinque parium, fuperioribus duarum triumve, omnibus obtufiufcule ferratis feu dentatis, dentibus apice purpureis.

LEAVES ftanding on foot-ftalks, pinnatifid, alternate, rough, the lowermoft confifting of four or five pair of pinnæ, the uppermoft of two or three, all of them obtufely ferrated or toothed, the teeth purple at the tips.

FLORES pedunculati, lutei, albi, feu carnei, venis nigricantibus picti.

FLOWERS ftanding on foot-ftalks, yellow, white or flefh-coloured, painted with blackifh veins.

CALYX: PERIANTHIUM tetraphyllum, erectum, hifpidulum, foliolis oblongis, parallelis, conniventibus, deciduis, bafi gibbis. *fig.* 1.

CALYX: a PERIANTHIUM of four leaves, upright, a little hifpid, the leaves oblong, parallel, clofing together, deciduous, gibbous at the bafe. *fig.* 1.

COROLLA tetrapetala, cruciformis: petala alba five lutea vel purpurafcentia, venofa: venis nigricantibus, obcordata, integra, patentia, fenfim in ungues calyce paulo longiores attenuata. *fig.* 2.

COROLLA compofed of four petals, which are white, yellow, or purplifh, veined (the veins blackifh) inverfely heart-fhaped, entire, fpreading, terminating gradually in claws, a little longer than the calyx. *fig.* 2.

STAMINA: FILAMENTA fex, fubulata, erecta; quorum duo oppofita longitudine calycis, quatuor vero unguibus longiora. ANTHERÆ oblongæ, erectæ, fagittatæ. *fig.* 3.

STAMINA: fix FILAMENTS, tapering, upright; of which two oppofite ones are of the length of the calyx, and four of the length of the claws of the corolla. ANTHERÆ oblong, upright, arrow-fhaped *fig.* 3.

PISTILLUM: GERMEN oblongum, ventricofum, attenuatum, longitudine ftaminum. STYLUS vix ullus. STIGMA capitatum, integrum. *fig.* 4.

PISTILLUM: GERMEN oblong, bellying out a little, tapering, the length of the ftamina. STYLE fcarce any. STIGMA forming a little head, and entire. *fig.* 4.

PERICARPIUM: Siliqua teres, articulata, articulis tribus ad fex et ultra, fulcatis, unilocularibus, monofpermis, geniculis attenuatis, apice roftrata, roftro lineari comprefto, glabra. *fig.* 5.

SEED-VESSEL a round jointed Pod, compofed of three, fix, or more joints, which are grooved, of one cavity, containing a fingle feed, the joints cut in, the apex terminating in a linear, flat, fmooth beak. *fig.* 5.

SEMINA fubrotunda, ferruginea, glaberrima, magnitudine feminum Raphani fativi. *fig.* 6.

SEEDS roundifh, ferrugineous, very fmooth, the fize of the common garden Radifh. *fig.* 6.

The prefent plant, in the colour of its bloffom, is one of the moft variable we are acquainted with, being found with white, pale-red, and yellow flowers: as the yellow is the moft predominant with us, we have figured that variety.

Though the bloffoms vary fo widely as to colour, they are all in general ftriped with purplifh veins; a character which contributes with feveral others to diftinguifh it from the *Sinapis Arvenfis*, or *Charlock*, to which it bears no fmall refemblance; fome of the moft ftriking differences between thefe two plants we fhall here enumerate.

The Charlock ufually grows one third taller than the Radifh. The ftalks, which in the Charlock are finely grooved, hirfute, and commonly of a deep red colour, in the Radifh are fmooth, yet hifpid, and ufually glaucous. The Charlock has often an unbranched ftem: the Radifh is more frequently branched quite down to the bottom; befides, the calyx is upright and clofe in the Radifh; in the Charlock it is fpreading. The flowers of the Charlock alfo are fmaller, and always yellow.

It is a common and noxious inhabitant of Corn-fields, and flowers in *July* and *Auguft*.

LINNÆUS has given a particular paper on this plant in the *Amœnitates Academicæ*, in which he endeavours to prove, that an epidemic fpafmodic difeafe, common in fome parts of *Sweden*, is owing to the feeds of this plant being ground with the corn and eaten by the inhabitants.

Turritis glabra.

Turritis glabra. Smooth Tower-Mustard.

TURRITIS *Lin. Gen. Pl.* TETRADYNAMIA SILIQUOSA.

Siliqua longiſſima, angulata. *Cal.* connivens, erectus. *Cor.* erecta.

Raii Syn. Gen. 21. HERBÆ TETRAPETALÆ, SILIQUOSÆ ET SILICULOSÆ.

TURRITIS *glabra* foliis radicalibus dentatis hiſpidis, caulinis integerrimis amplexicaulibus glabris. *Lin. Syſt. Vegetab.* p. 502. *Sp. Pl.* p. 930. *Haller. Hiſt.* n. 455.

TURRITIS *glabra. Scopoli Fl. Carn.* n. 839.

BRASSICA ſylveſtris foliis circa radicem cichoracea. *Baub. Pin.* 112.

TURRITIS foliis inferioribus cichoraceis, cæteris perfoliatis. *J. R. H.*

BRASSICA ſylveſtris ramoſa tota penè glabra. *Baub. Pin.* 112.

TURRITIS vulgatior. *Baub. Hiſt.* II. 836.

TURRITIS *Ger. emac.* 272. *Parkins* 852. *Raii Syn.* p. 293. Tower-Muſtard. *Hudſon Fl. Angl.* ed. 2. p. 291.

RADIX biennis, ſimplex, alba, ſublignoſa, alte in terram, deſcendens, paucis fibrillis capillata.

CAULIS pedalis ad tripedalem, erectus, plerumque ſimplex, firmus, teres, ſolidus, prope terram hirſutus, ſupernè glaber.

FOLIA radicalia pallide viridia, hirſuta, ad margines ſinuata, tres quatuorve uncias longa, caulina glauca, glabra, integerrima, amplexicaulia, erecta, ſagittata.

FLORES in ſummis caulibus parvi, ex albo luteſcentes.

CALYX: PERIANTHIUM tetraphyllum, foliolis ovato-oblongis, erectis, deciduis. *fig.* 1.

COROLLA: tetrapetala, cruciformis. *Petala* ovato-oblonga, obtuſa, integra, unguibus erectis. *fig.* 2.

STAMINA: FILAMENTA ſex, ſubulata, alba, quorum duo breviora. ANTHERÆ ſimplices, flavæ. *fig.* 3.

PISTILLUM: GERMEN longitudine floris, teres, ſubcompreſſum. STYLUS nullus. STIGMA obtuſum. *fig.* 4.

PERICARPIUM: *Siliquæ* plurimæ, pedicellatæ, erectæ, duos vel tres digitos longæ, caulem fere occultantes, teretiuſculæ, ſubcompreſſæ, obſoletè quadrangulæ. *fig.* 5.

SEMINA parva, numeroſiſſima, ovata, planiuſcula, rufa. *fig.* 6.

ROOT biennial, ſimple, white, ſomewhat woody, penetrating deeply into the earth, furniſhed with few fibres.

STALK from one to three feet in height, upright, generally ſimple, firm, round, ſolid, near the ground hirſute, above ſmooth.

LEAVES next the root, of a pale green colour, ſtrongly hairy, jagged on each ſide, three or four inches in length, thoſe of the ſtalk glaucous, perfectly ſmooth and entire, embracing the ſtalk, upright, and arrow-ſhaped.

FLOWERS on the top of the ſtalks, ſmall, of a whitiſh yellow colour.

CALYX: a PERIANTHIUM of four leaves, the leaves of an ovate oblong ſhape, upright and deciduous. *fig.* 1.

COROLLA tetrapetalous and croſs-ſhaped. *Petals* of an ovate-oblong ſhape, obtuſe, entire, claws upright. *fig.* 2.

STAMINA: ſix FILAMENTS, tapering, white, two of which are ſhorter than the others. ANTHERÆ ſimple, and yellow. *fig.* 3.

PISTILLUM: GERMEN the length of the flower, round, a little flattened. STYLE none. STIGMA blunt. *fig.* 4.

SEED-VESSEL: *Pods* numerous, ſtanding on footſtalks, upright, two or three fingers breadth in length, almoſt hiding the ſtalk, roundiſh, ſomewhat flattened, faintly quadrangular. *fig.* 5.

SEEDS ſmall, very numerous, ovate, flattiſh, of a reddiſh brown colour. *fig.* 6.

The term *glabra* is only proper when applied to the upper part of this plant, the radical leaves, which generally decay as the plant approaches to maturity, being hairy, like many other plants of the ſame family.

It is found but rarely near *London*. Hitherto I have only noticed it in one ſpot, *viz.* in the lane which leads down by *Charlton* Church, and there but ſparingly; further on in *Kent* it is much more common, as well as in many other parts of *England*. It uſually grows on banks near hedges, and flowers in *June* and *July*.

It varies ſo much in ſize, that the old Botaniſts make two ſpecies of it.

CARDAMINE HIRSUTA. HAIRY LADIES-SMOCK.

CARDAMINE *Lin. Gen. Pl.* TETRADYNAMIA SILIQUOSA.

> *Siliqua* elaſtice diſſiliens valvulis revolutis. *Stigma* integrum. *Cal.* ſubhians.

Raii Syn. Gen. 21. HERBÆ TETRAPETALÆ SILIQUOSÆ ET SILICULOSÆ.

CARDAMINE foliis pinnatis, floribus tetrandris. *Lin. Syſt. Vegetab.* p. 497. *Sp. Pl.* p. 915. *Fl. Suec.* n. 587.

CARDAMINE foliis pinnatis hirſutis, pinnis ſubrotundis, ſtaminibus quaternis. *Haller Hiſt.* 472.

CARDAMINE *hirſuta. Scopoli Fl. Carn.* n. 817. t. 38.

NASTURTIUM aquaticum minus. *Bauh. Pin.* 104.

CARDAMINE impatiens altera hirſutior. *Raii Syn.* p. 300. The leſſer hairy, impatient Cuckow-flower or Ladies ſmock.

CARDAMINE foliis pinnatis, foliolis radicalibus ſubrotundo-cordatis; caulinis ovatis dentatis petiolatis. *Hudſon. Fl. Angl. ed.* 2. p. 295. *Lightfoot Fl. Scot.* p. 348.

RADIX annua, fibroſa, fibris albidis. ‡ ROOT annual and fibrous, the fibres whitiſh.

CAULIS ſpithamæus, et ultra pro ratione loci, in foſſis humidis reperitur etiam ſeſquipedalis, ſolidus, erectus, flexuoſus, ſulcato-anguloſus, prope baſin purpureus, et ſæpius hirſutiſſimus, ſupente fere glaber, ramoſus, ramoſiſſimus etiam occurrit. ‡ STALK about a ſpan high, or more, according to the ſituation in which it grows; in wet ditches it is ſometimes found even a foot and a half in height, ſolid, upright, crooked, grooved or angular, purple near the baſe, and moſt commonly very hairy, above hearly ſmooth, branched, ſometimes very much ſo.

FOLIA radicalia plurima, in orbem poſita, pinnata, foliolis petiolatis, rotundato-angulatis, plerumque quinque lobatis, hirſutis, punctis prominulis ſcabriuſculis, lobis inæqualibus, nunc obtuſis, nunc acutis; caulina anguſtiora et magis profunde inciſa, lobis paucioribus. ‡ LEAVES next the root numerous, forming a circle, pinnated, the ſmall leaves ſtanding on footſtalks, round yet angular, generally divided into five lobes, hirſute, roughiſh with little prominent points; the lobes unequal, ſometimes blunt and ſometimes pointed; thoſe of the ſtalk narrower, and more deeply indented, with fewer lobes.

FLORES parvi, albi, primo vere tantum, tetrandri. ‡ FLOWERS ſmall, and white, early in the ſpring, having only four ſtamina.

CALYX: PERIANTHIUM tetraphyllum, foliolis ovato-oblongis, obtuſis, concavis, deciduis, pilis paucis albidis inſtructis. *fig.* 1. ‡ CALYX: a PERIANTHIUM of four leaves, of an ovate, oblong ſhape, obtuſe, hollow, deciduous, furniſhed with a few white hairs. *fig.* 1.

COROLLA: PETALA quatuor, alba, calyce duplo fere longiora, patentia, integerrima, obtuſa. *fig.* 2. ‡ COROLLA: four white PETALS, almoſt twice the length of the calyx, ſpreading, entire and obtuſe. *fig.* 2.

STAMINA: FILAMENTA plerumque ſex, quorum duo breviora, alba. ANTHERÆ minimæ, luteſcentes. *fig.* 3. ‡ STAMINA: for the moſt part ſix FILAMENTS, of which two are ſhorter than the reſt, of a white colour. ANTHERÆ very ſmall and yellowiſh. *fig.* 3.

PISTILLUM: GERMEN oblongum, tenue, ſtaminibus paulo brevior. STIGMA capitatum. *fig.* 4. ‡ PISTILLUM: GERMEN oblong, ſlender, a little ſhorter than the ſtamina. STIGMA forming a ſmall head. *fig.* 4.

PERICARPIUM: *Siliqua* erecta, uncialis, ſubcompreſſa, bivalvis, elaſtice diſſiliens, valvulis revolutis. *fig.* 5. ‡ SEED-VESSEL: an upright *Pod*, about an inch in length, ſomewhat flattened, of two valves, which burſt with an elaſtic force, and roll back. *fig.* 5.

SEMINA duodecim circiter, ſuborbiculata, compreſſa, glabra, e flavo fuſca. *fig.* 6. ‡ SEEDS about twelve in number, nearly round and flattened, ſmooth, and of a yellowiſh-brown colour. *fig.* 6.

We were inclined to believe with our ingenious friend Mr. LIGHTFOOT, that the *Cardamine hirſuta* and *parviflora* were diſtinct ſpecies; but repeated obſervation and culture have convinced us, that they are both the ſame, varying only in ſize, in hairineſs, and in the number of their ſtamina.

In wet ſituations, where the ſoil is luxuriant, it grows a foot or two in height, and loſes in a great degree its hairineſs; in expoſed places it ſeldom reaches more than ſix or eight inches, and is generally much more hairy, and, when it grows ſingly, much more branched. The ſame plant, early in the ſpring, when the weather is cold, has only four ſtamina; as the ſumer advances, it has conſtantly ſix. The lobes of the radical leaves vary much in ſhape, and are frequently much rounder than the figure repreſents.

This ſpecies is by no means general about London, but abounds in particular places; as by Chelſea water-works, in wet ditches about Hampſtead, Highgate, and elſewhere.

It flowers in April and May. In the garden, if the ſituation in which it is ſown be ſhady, and the ſeaſon not uncommonly dry, it continues flowering and ſeeding during the whole of the ſummer.

According to Mr. LIGHTFOOT, the young leaves are a good ingredient in a ſallad, and may eaſily be obtained in the ſpring, when Muſtard and Creſs are not to be had.

Cardamine hirsuta.

GERANIUM PRATENSE. CROWFOOT CRANESBILL.

GERANIUM *Lin. Gen. Pl.* MONADELPHIA DECANDRIA.

  Monogyna. *Stigmat.* 5. *Fructus* rostratus, 5-coccus.

 Raii Syn. Gen. 24. HERBÆ PENTAPETALÆ VASCULIFERÆ.

GERANIUM *pratense* pedunculis bifloris, foliis fubpeltatis multipartitis rugofis acutis, petalis integris. *Lin. Syst. Veget.* p. 514. *Sp. Pl.* p. 954. *Fl. Suec.* n. 968.

GERANIUM caule erecto, foliis rugofis hirfutis multilobis, lobis trifidis, lobulis femipinnatis, floribus umbellatis. *Haller. Hist.* n. 931.

GERANIUM *pratense Scopoli Fl. Carn.* n. 852.

GERANIUM batrachoides, Gratia Dei Germanorum. *Baub. Pin.* 318.

GERANIUM Batrachoides. *Ger. emac.* 922.

GERANIUM Batrachoides flore cærulea. *Park.* 704. *Raii Syn.* ed. 3. p. 360. Crowfoot Cranefbill. *Hudfon Fl. Angl.* ed. 2. p. 302. *Lightfoot Fl. Scot.* p. 368.

RADIX perennis, craffitie digiti minimi aut major, horizontalis, rugofa, e rubro fufca, intus flavefcens, fibris majufculis profunde penetrantibus inftructa.

ROOT perennial, the thicknefs of the little finger, or larger, horizontal, wrinkled, of a reddifh brown colour, yellowifh within, furnifhed with largifh fibres, which penetrate to a great depth.

CAULIS pedalis ad tripedalem, erectus, ramofus, teretiufculus, pubefcens, bafi rubertimus, fubangulofus.

STALK from one to three feet in height, upright, branched, roundifh, downy, very red, and fomewhat angular at the bottom.

FOLIA hirfutula, radicalia longiffime petiolata, fuprema fubfeffilia, omnibus multipartitis, venofis, fubtus pallidioribus, laciniis multifidis, incifis.

LEAVES fomewhat hirfute, thofe next the root, ftanding on very long footftalks, the uppermoft ones nearly feffile, all of them deeply divided into many fegments, veiny and paler underneath, the fegments jagged.

PETIOLI teretes, pubefcentes.

LEAF-STALKS round and downy.

STIPULÆ ad bafin foliorum utrinque binæ, ovato acuminatæ, primo rubicundæ, dein marcefcentes, ad bafin pedunculorum quinæ, lanceolatæ.

STIPULÆ at the bafe of the leaves two on each fide, ovate and pointed, at firft reddifh, afterwards withering, thofe at the bafe of the peduncles five in number, and lanceolate.

PEDUNCULI gemini, pilofiffimi, vifcofi, primo nutantes, demum erecti.

FLOWER-STALKS growing two together, very hairy, clammy, at firft drooping, laftly upright.

FLORES magni, fpeciofi, e purpureo cærulei.

FLOWERS large, fhowy, of a purplifh blue colour.

CALYX: PERIANTHIUM pentaphyllum, perfiftens, foliolis ovatis, acutis, concavis, margine membranaceis. *fig.* 1.

CALYX: a PERIANTHIUM of five leaves and permanent, the leaves ovate, pointed, concave, bearded, clammy, ribbed and membranous at the edge. *fig.* 1.

COROLLA: PETALA quinque, fubintegerrima, venofa, bafi utrinque hirfutula. *fig.* 2.

COROLLA: five PETALS, nearly entire, veiny, hairy on each fide the bafe. *fig.* 2.

STAMINA: FILAMENTA decem fubulata, fuperne purpurafcentia, inferne lata, albida; ANTHERÆ oblongæ, cæruleæ, incumbentes. *fig.* 3.

STAMINA: ten FILAMENTS, tapering, purplifh above, below broad and whitifh; ANTHERÆ oblong, blue and incumbent. *fig.* 3.

PISTILLUM: GERMEN pentagonum, vifcofum, viride; STYLUS filiformis, rubicundus, ftaminibus longior; STIGMA quinquefidum, laciniis linearibus, reflexis. *fig.* 4.

PISTILLUM: GERMEN pentagonal, clammy, of a green colour; STYLE filiform, reddifh, longer than the ftamina; STIGMA divided into five fegments, which are linear and turned back. *fig.* 4.

SEMEN oblongum, læve, ex arillo elaftice diffiliens. *fig.* 5, 6.

SEED oblong, fmooth, thrown from the feed-covering with confiderable elafticity. *fig.* 5, 6.

 This is by far the moft fhowy of the Craneftalls growing wild with us, and in that refpect is exceeded by none of the Englifh ones except the *Sanguineum*. Its leaves are large, and fomewhat like thofe of the Meadow Crowfoots, whence its name.

 Its beauty has long fince recommended it to the notice of the curious in flowers, in the gardens of which we find it not only as it commonly occurs, but alfo with white, purple, and ftriped bloffoms.

 It loves a moift fituation, as it naturally grows wild in meadows, which it ornaments with its bloffoms in July and Auguft. Near London it is found in tolerable abundance in the meadows about Batterfea, and in the vicinity of the *Thames* both above and below bridge; in many parts of *Yorkfhire*, particularly about *Settle*, it is almoft as common as the Mallow is with us, not only in meadows, but every where under their ftone walls.

 The bloffoms are much reforted to by various fpecies of flies, particularly thofe of the Genus *Empis*.

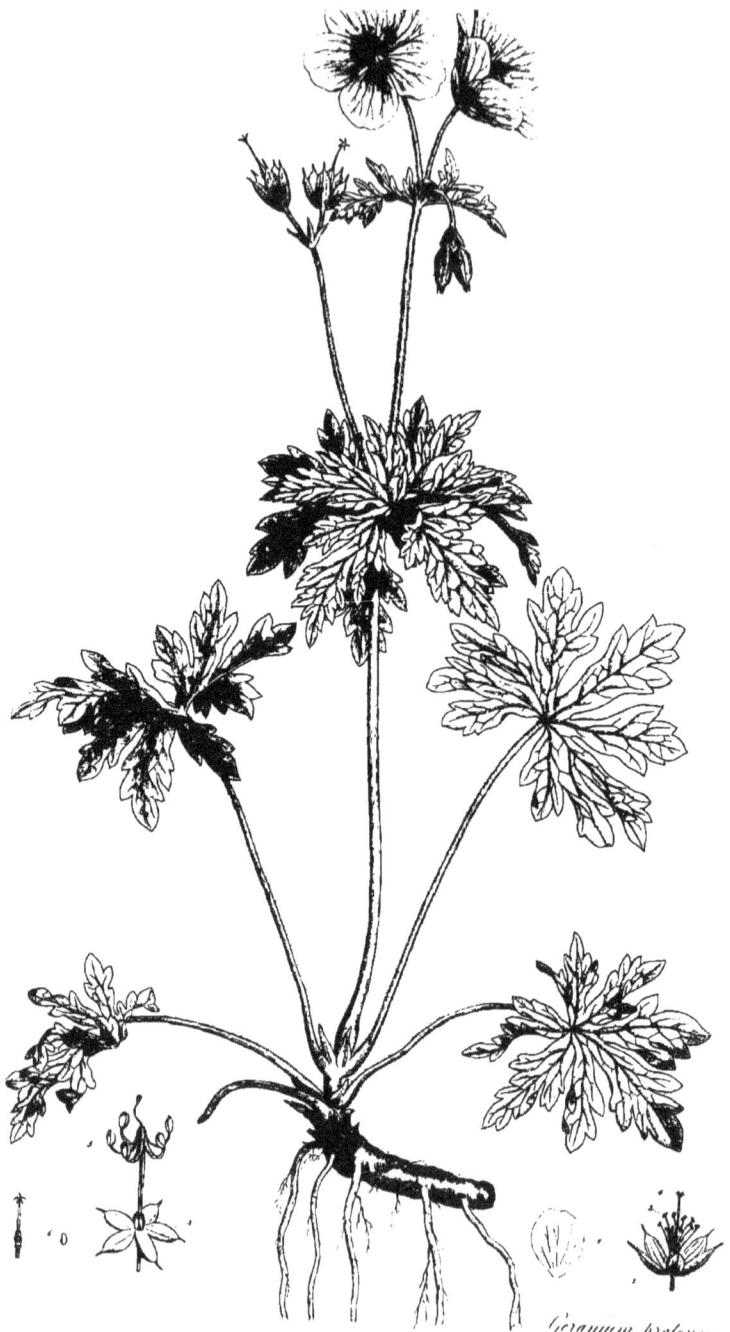

Géranium pratense

MALVA MOSCHATA. MUSK MALLOW.

MALVA *Lin. Gen. Pl.* MONADELPHIA POLYANDRIA.
    Cal. duplex: exterior triphyllus. *Arilli* plurimi, monospermi.
    Raii Syn. Gen. 15. HERBÆ SEMINE NUDO POLYSPERMÆ.
MALVA *moschata* caule erecto, foliis radicalibus reniformibus incisis; caulinis quinquepartitis pinnato-
    multifidis. *Lin. Syst. Vegetab.* p. 523. *Spec. Pl.* p. 971. *Fl. Suec.* n. 629.
MALVA foliis radicalibus reniformibus, incisis, caulinis quinquepartitis pinnatis, pinnis dentatis.
    Haller. Hist. n. 1072.
MALVA *Moschata. Scopoli Fl. Carn.* n. 861.
MALVA montana five Alcea rotundifolia laciniata. *Col. Ecphr.* 1. p. 148. t. 147.
ALCEA folio rotundo laciniato. *Bauh. Pin.* 316.
ALCEA tenuifolia crispa. 1. B. II. *App.* 1067. *Raii Syn.* p. 253. *Hudson. Fl. Angl. ed.* 2. p. 308.
    Lightfoot Fl. Scot. p. 376.

RADIX perennis, albida, sublignosa, difficillime eruta.

CAULIS: ex una radice caules nascuntur plurimi, bipedales, suberecti, ramosi, teretes, fistulosi, hirsuti, punctis purpureis prominulis adspersi, e
quibus pili prodeunt.

FOLIA alterna, petiolata, inferiora plerumque sex partita, laciniis pinnatifidis, multifido-laciniatis,
hirsutulis, apice subacutis, superiora brevius
petiolata, in pauciores et tenuiores lacinias
divisa.

STIPULÆ utrinque binæ, erectæ, lanceolatæ, hirsutæ,
marginæ undulatæ.

FLORES magni, speciosi, carnei.

PEDUNCULI unciales, teretes, pilosi.

CALYX: PERIANTHIUM duplex, inferius triphyllum,
foliolis lanceolatis, superius quinquefidum,
ovato-acutum, sæpe laciniatum, punctatum,
hirsutum, margine ferrato glandulofum. *fig.* 1.

COROLLA: PETALA quinque, subtriangularia, carnea,
venis saturationibus ramosis notata, apice sub-
truncata, erosa, basi albida, margine utrinque
ciliata. *fig.* 2.

STAMINA: FILAMENTA plurima, in tubum cylindra-
ceum, albidum, pilosum coalita, superne libera, reflexa. ANTHERÆ primum reniformes,
carneæ, dein purpureæ, demum cæruleicentes.
POLLEN album, globolum. *fig.* 3.

PISTILLUM: GERMINA plurima, in orbem disposita,
flavescentia; STYLI plurimi, ruberrimi, filiformes, ad unum latus pilosi, polline plerumque obducti. STIGMATA simplicia. *fig.* 5.

ROOT perennial, whitish, somewhat woody, and with
difficulty pulled up.

STALK: from one root arise several stalks, about two
feet high, nearly upright, branched, round,
hollow, hirsute, sprinkled with purple prominent points, from which the hairs issue.

LEAVES alternate, standing on foot-stalks, the lower
ones generally deeply divided into six segments,
which are pinnatifid and sub-divided into many
others, slightly hirsute, and pointed at the
tips, the upper ones standing on shorter foot-
stalks, and divided into fewer and narrower
segments.

STIPULÆ two on each side, upright, lanceolate, hirsute, waved on the edge.

FLOWERS large, showy, and flesh-coloured.

FLOWER-STALKS an inch in length, round and
hairy.

CALYX: a double PERIANTHIUM, the lowermost composed of three lanceolate leaves, the uppermost divided into five segments, ovate and
pointed, often jagged, dotted, hirsute, the
edge ferrated with small glands. *fig.* 1.

COROLLA: five PETALS, somewhat triangular, flesh-
coloured, marked with branched veins of a
deeper colour, somewhat truncated with a
piece bit out at top, at bottom whitish, with
the edge fringed on each side with hairs. *fig.* 2.

STAMINA: FILAMENTS numerous, forming a whitish
hairy cylindrical tube, loose at top, and bending back. ANTHERÆ at first kidney-shaped
and flesh-coloured, then purple, and lastly
blueish. POLLEN white and globular. *fig.* 3.

PISTILLUM: GERMINA numerous, circularly disposed,
of a yellowish colour; STYLES numerous, of
a bright red colour, thread-shaped, hairy on
one side, and generally covered with pollen.
STIGMATA simple. *fig.* 5.

The plant here figured has been, and is still, considered by most English Botanists as the *Vervain Mallow:* there
is little doubt but it is the plant which RAY considered as the *Alcea vulgaris major* of C. BAUHINE, in which idea
he was most probably mistaken, as it accords better with the *Alcea rotundifolia laciniata* of that author. Be this
as it may, it certainly is not the *Malva Alcea* of LINNÆUS, which Mr. HUDSON makes a native of this country;
and which, he says, grows wild in *Warwickshire, Leicestershire,* and *Nottinghamshire.*

At the same time that LINNÆUS has very properly made two distinct species of these plants, he has been rather
unfortunate in the parts he has selected for their discrimination.

Taking for granted that Mr. HUDSON has good authority for what he afferts (although the counties he specifies are
not particularly mentioned by RAY) it appears, that the *Moschata* is a general, the *Alcea* a local plant; the
former I have found in most of the counties I have visited, and Mr. LIGHTFOOT mentions it as growing in
Scotland; the latter I have never seen wild, but have occasionally observed it in some of the gardens about London;
and last year, having an opportunity of cultivating it in my own, I was agreeably surprized, to find that it afforded
many obvious and satisfactory distinctions, the most striking of which I shall enumerate, for the gratification of
the English Botanist.

The *Malva Alcea* grew to nearly twice the height of the *Moschata,* whence it agrees with BAUHINE's name of
major. It was in every respect a stronger plant, and harsher to the touch; the leaves of the stalk were much less
jagged; the flowers in both were pretty similar, both in shape, size, and colour; but the calyces differed
remarkably. In the *Moschata* the lowermost set of leaves were lanceolate, sometimes almost linear; in the
Alcea they were ovate; added to this, the calyx of the *Alcea,* near its base, had a large protuberant annulus
or ring, which was entirely wanting in the *Moschata.* These characters of the calyx alone will, it is presumed,
ever be found sufficient to distinguish the two plants when in flower; besides these, the *Moschata* drawn through
the hand has the peculiar property of communicating a strong smell of musk, whence its name.

From this relation those Botanists, resident in the counties above mentioned, will be better enabled to judge
whether they have the true *Malva Alcea* or not.

The *Moschata* grows very plentifully in the neighbourhood of *Coomb Wood,* and flowers in *June* and *July.*

No particular virtues or uses are attributed to this species; but its beauty entitles it to a place in the garden.

Bees resort much to it.

Malva moschata.

Trifolium Glomeratum.

TRIFOLIUM *Lin. Gen. Pl.* Diadelphia Decandria.

> *Flores* fubcapitati. *Legumen* vix calyce longius, non dehifcens, deciduum.

Raii Syn. Gen. 23. Herbæ flore papilionaceo seu leguminosæ.

TRIFOLIUM *glomeratum* capitulis feffilibus hemifphæricis rigidis, calycibus ftriatis patulis æqualibus. *Lin. Syft. Vegetab.* p. 573. *Sp. Plant.* p. 1084.

TRIFOLIUM arvenfe fupinum verticillatum. *Barr. ic.* 882.

TRIFOLIUM parvum rectum, flore glomerato cum unguiculis. *J. B.* II. 378.

TRIFOLIUM cum glomerulis ad caulium nodos rotundis. *Raii Syn. ed.* 3. p. 329. Knotted Trefoil, with round heads. *Hudfon. Fl. Angl. ed.* 2. p. 327.

RADIX annua, fimplex, fibrofa.

CAULES plures, palmares, procumbentes, teretes, purpurafcentes, glabri, ramofi, geniculati, geniculis tumidiufculis.

STIPULÆ caulinæ membranaceæ, acuminatæ, ramulorum apice fæpius recurvis.

FOLIA glabra, viridia, macula alba fæpius notata, minute ferrata.

FLORES glomerati, rubelli, axillares, capitulis feffilibus, hemifphæricis.

CALYX: Perianthium quinquedentatum, perfiftens, læve, ftriis decem coloratis notatum, dentibus ovato-acuminatis, patulis. *fig.* 1.

COROLLA longitudine calycis ; *vexillum* furfum curvatum ; *alæ* vexillo duplo breviores, apice paululum fecedentes ; *carina* longitudine fere alarum. *fig.* 2.

PERICARPIUM: Legumen membranaceum, difpermum, intra calycem. *fig.* 3.

SEMINA minima, flavefcentia. *fig.* 4.

ROOT annual, fimple and fibrous.

STALKS feveral from the fame root, four or fix inches in length, procumbent, round, purplifh, fmooth, branched and jointed ; the joints a little fwelled.

STIPULÆ of the ftalk membranous and pointed, thofe of the branches often bent back at top.

LEAVES fmooth, green, often marked with a white fpot, the edge finely fawed.

FLOWERS cluftered, of a pink colour, growing in the alæ of the leaves, the little heads feffile, and almoft globular.

CALYX: a Perianthium having five teeth, permanent, fmooth, marked with ten coloured ftreaks, the teeth broad, pointed, and fpreading. *fig.* 1.

COROLLA the length of the calyx ; *ftandard* bent upwards ; *wings* about half as long as the ftandard, feparating a little at their extremities ; *keel* almoft the length of the wings. *fig.* 2.

SEED-VESSEL : a membranous Pod, containing two feeds within the calyx. *fig.* 3.

SEEDS very minute, and yellowifh. *fig.* 4.

The bloffoms of this fpecies of *Trefoil* grow in little round balls or clufters from the alæ of the leaves, and hence it has received its name of *glomeratum*.

It may be diftinguifhed from the *ftriatum*, to which it bears fome affinity, by being in every part fmooth, in having its balls much rounder, and the teeth of the calyx fpreading backward; its bloffoms alfo are of a brighter red colour.

Not being a plant which ftrikes the eye much at a diftance, it is probably often overlooked ; with us, however, it is certainly fcarce. Mr. Ray found it about *Saxmundham* in *Suffolk* ; Mr. Doody about *Blackheath*, and near *Greenbithe* ; Mr. Hudson in the *Ifle of Shepey* ; Mr. Rose about *Norwich*, and Dr. Goodenough on *Hanwell Heath*. I have found it on *Blackheath* for feveral years, yet not plentifully ; it is fond of a gravelly fituation, with fome degree of moifture ; hence it muft be looked for in the depreffed parts of the heath. The plants growing near it were the *Trifolium ftriatum, ornithopodioides*, and *Sagina procumbens* ; and at no great diftance *Callitriche autumnalis, Montia fontana*, and *Peplis Portula*.

It flowers in June.

Hypericum quadrangulum

HYPERICUM QUADRANGULUM. SQUARE-STALKED ST. JOHN'S WORT.

HYPERICUM *Lin. Gen. Pl.* POLYADELPHIA POLYANDRIA.

> *Cal.* 5-partitus, *Petala* 5, *Filamenta* multa in 5 phalanges bafi coanata. *Capfula.*

Raii Syn. Gen. 24. HERBÆ PENTAPETALÆ VASCULIFERÆ.

HYPERICUM *quadrangulum* floribus trigynis, caule quadrato herbacea. *Lin. Syfl. l'egetab. p.* 584. *Sp. Pl. p.* 1104. *Fl. Suec.* n. 670.

HYPERICUM caule quadrangulari, foliis ovatis perforatis punctatis. *Haller. Hift.* n. 1038.

HYPERICUM *quadrangulum. Scopoli Fl. Carn.* n. 943.

HYPERICUM vulgare minus caule quadrangulo foliis non perforatis. *Bauh. pin.* 272.

HYPERICUM Afcyron dictum caule quadrangulo. *Bauh. Hift.* 3. p. 382.

ASCYRON *Dod. pempt.* 78. *Ger. emac.* 542. *vulgare Parkinfon* 575. *Raii Syn.* p. 344. St. Peter's Wort. *Hudfon. Fl. Angl. ed.* 2. p. 334. *Lightfoot Fl. Scot.* p. 416.

RADIX perennis, fubrepens, fibrofa.

CAULIS pedalis ad fefquipedalem, erectus, ramofus, glaber, rubefcens, quadratus, quatuor membranulis, feu alis in fecundum longitudinem productis.

FOLIA oppofita, feffilia, ovata, obtufa, lævia, faturate viridia, fubtus pallidiora, feptemnervia, per totam fuperficiem punctis minutis diaphanis adfperfa, margine glandulis nigris fubtus præfertim punctata.

RAMI cauli fimiles, decuffatim oppofiti.

FLORES lutei, parvi, in fummitatibus ramulorum dente paniculati.

RAMULI paniculæ fanguinei.

PEDUNCULI breviffimi.

CALYX: PERIANTHIUM quinquepartitum, laciniis lanceolatis, patentibus, nervofis, integerrimis, nudis. *fig.* 1.

COROLLA: PETALA quinque, lutea, lineata, glandulis paucis nigris ad oras punctata. *fig.* 2.

STAMINA: FILAMENTA plurima, in fafciculos vix divifa; ANTHERÆ fubrotundæ, flavæ, glandulâ nigrâ notatæ. *fig.* 3.

PISTILLUM: GERMEN obovatum; STYLI tres, patentes, fubulati; STIGMATA minima, capitata. *fig.* 4.

PERICARPIUM: CAPSULA fufca, trilocularis, trivalvis. *fig.* 5.

SEMINA plurima, minima, oblonga, nitidiufcula. *fig.* 6.

ROOT perennial, fomewhat creeping and fibrous.

STALK from a foot to a foot and a half in height, upright, branched, fmooth, reddifh, fquare from having four little membranes or wings which run down the ftalk.

LEAVES oppofite, feffile, ovate, obtufe, fmooth, of a deep green colour, paler underneath, ftrongly marked with feven ribs, the whole furface covered with fmall tranfparent dots, and the edge, efpecially on the under fide, dotted with black glands.

BRANCHES like the ftalk, alternately oppofite.

FLOWERS of a yellow colour, fm.ll, growing on the tops of the branches in clofe panicles.

BRANCHES of the panicle of a deep red or blood colour.

FLOWER-STALKS very fhort.

CALYX: a PERIANTHIUM deeply divided into five fegments, which are lanceolate, fpreading, rib'd, entire, and free from glands. *fig.* 1.

COROLLA: five yellow PETALS, finely grooved, dotted on the edge with a few fmall black glands. *fig* 2.

STAMINA: FILAMENTS numerous, fcarcely divided into bodies or bundles; ANTHERÆ roundifh, yellow, marked with a black gland. *fig.* 3.

PISTILLUM: GERMEN inverfely ovate; STYLES three, fpreading, tapering; STIGMATA very fmall, forming little heads. *fig.* 4.

SEED-VESSEL: a brown CAPSULE of three cavities and three valves. *fig.* 5.

SEEDS numerous, very fmall, oblong, and fomewhat fhining. *fig.* 6.

The *Saint-John's-Worts* (at leaft of this country) are a genus of plants which, having ftrong characteriftic marks, and being fubject to little variation, give the Botanift no great trouble in their inveftigation; it is fufficient to fay of this fpecies for inftance, that it has a fquare ftalk, and it is at once diftinguifhed from all the others.

CASPAR BAUHINE commits no fmall error when he defcribes the leaves of the *quadrangulum* as imperforate, the leaft attention will fhew the tranfparent dots on the leaves to be fully as numerous, if not fo large as thofe of the *perforatum*; the antient Botanifts alfo abfurdly enough diftinguifhed this fpecies from the others by the name of *Afcyron*, or *Saint-Peter's-Wort*, but as it has no pretenfions to any generic diftinction, we have dropped that name, as tending much to confufe: and while we are cenfuring the faults of others, we fhall mention an error of our own; in defcribing the *Hypericum perforatum* we pointed out a little black gland vifible betwixt the lobes of the antheræ, as characteriftic of that fpecies, we now find the fame on the *quadrangulum* alfo.

This fpecies grows very plentifully by the fides of rivulets, alfo in wet meadows. By the fides of the ditches in *Battterfea Meadows* it is particularly common, and flowers in July.

It is feldom ufed in medicine, the *perforatum* fupplying its place.

Sonchus arvensis.

SONCHUS ARVENSIS: CORN SOW-THISTLE.

SONCHUS *Lin. Gen. Pl.* SYNGENESIA POLYGAMIA ÆQUALIS.

Recept. nudum. *Cal.* imbricatus, ventricofus. *Pappus* pilofus.

Raii Syn. Gen. 5. HERBÆ FLORE COMPOSITO NATURA PLENO LACTESCENTES.

SONCHUS *arvenfis* pedunculis calycibufque hifpidis fubumbellatis, foliis runcinatis bafi cordatis. *Lin. Syft. Vegetab.* p. 594. *Spec. Pl.* 1116. *Fl. Succ.* n. 687.

SONCHUS foliis amplexicaulibus femipinnatis ferratis, calycibus hifpidis. *Haller. Hift.* 23.

HIERACIUM arvenfe. *Scopoli Fl. Carniol.* n. 974.

HIERACIUM majus folio fonchi. *Bauhin. Pin.* 126.

SONCHUS repens multis Hieracium majus. *Bauh. Hift.* 2. 176.

SONCHUS arborefcens. *Ger. emac.* 294. *Raii Syn.* p. 163. Tree Sow-thiftle. *Hudfon Fl. Angl.* ed. 2. p. 337. *Lightfoot Fl. Scot.* 427.

RADIX *perennis*, lactiflua, *longe, lateque repens*, difficulter eruta.

ROOT *perennial*, milky, *creeping far and wide*, with difficulty eradicated.

CAULIS bi feu tripedalis et ultra, erectus, fiftulofus, lactefcens, fubangulatus, lævis, inferne fæpe purpureus, fuperne ramofus.

STALK from two to three feet high, or more, upright, hollow, milky, fomewhat angular, fmooth, often purple below, above branched.

FOLIA alterna, runcinata, *bafi cordata*, amplexicaulia, lævia, nitidula, fubtus pallidiora, fpinis mollicellis circa margines donata.

LEAVES alternate, like thofe of Dandelion, *heart fhaped at the bafe*, embracing the ftalk, fmooth, fhining, paler on the under fide, the edges furnifhed with foftifh prickles.

FLORES fubumbellati, *magni*, lutei.

FLOWERS growing in a kind of umbel, *large* and yellow.

PEDUNCULI longi, teretes, *pilis longis luteis*, globuligeris veftiti.

FLOWER-STALKS long, round, covered with long, yellow, hairs, bearing globules at their extremity.

CALYX communis priufquam flofculi expanduntur cylindricus, apice truncatus, poftea ventricofoconicus, fquamis plurimis, erectis, inæqualibus, carinatis, fordide viridibus, *hirfutiffimis*, pilis ficut in pedunculis.

CALYX common to all the florets, before their expanfion, cylindrical and cut off at the extremity, afterwards bellying at the bafe and conical, the fcales numerous, upright, unequal, keeled, of a dirty green colour, and extremely hairy, the hairs fimilar to thofe on the footftalks.

COROLLA compofita, imbricata, *Corollulis* hermaphroditis, numerofis, æqualibus.

COROLLA compound and imbricated, the Florets hermaphrodite, numerous and equal.

Propria monopetala, *tubus* albus, tenuis, fuperne pilofus, limbus luteus, longitudine fere tubi marginibus fæpe fubinvolutis, quinquedentatus. *fig.* 1.

Each Floret monopetalous, the *tube* white, flender, hairy above, the flat part yellow, almoft the length of the tube, the edges frequently rolled inward, furnifhed with five teeth. *fig.* 1.

ANTHERÆ in tubum flavum, fubangulatum, coalitæ.

ANTHERÆ united into a yellow, and fomewhat angular tube.

STIGMATA duo, filiformia, reflexa.

STIGMATA two, thread-fhaped, reflexed.

SEMEN ovato-oblongum, badium, fulcatum, angulis ad lentem tranfverfim rugofis, pappofum, pappo fimplici, feffili. *fig.* 2.

SEED of an oblong egg-fhape, bay colour, and grooved, the angles tranfverfely wrinkled when magnified, downy, the down fimple, and feffile. *fig.* 2.

This fpecies of *Sonchus* is properly termed *arvenfis*, being commonly found in corn fields, in which its large yellow bloffoms, towering above the corn, render it a very confpicuous plant; thefe alone are fufficient to diftinguifh it from the common Sow-thiftle, it has befides two other very diftinct characters; the one a creeping root, whereby it becomes very noxious to the hufbandman; the other, numerous yellow hairs, with little globules at their extremities, thick fpread over the calyces and flower-ftalks.

It bloffoms in *July* and *Auguft*; many of its feeds prove abortive.

Hieracium Pilosella

HIERACIUM PILOSELLA. MOUSE-EAR.

HIERACIUM *Lin. Gen. Pl.* SYNGENESIA POLYGAMIA ÆQUALIS.

Recept. nudum. *Cal.* imbricatus, ovatus. *Pappus* simplex, sessilis:

Raii Syn. Gen. 6. HERBÆ FLORE COMPOSITO NATURA PLENO LACTESCENTES.

HIERACIUM *Pilosella* foliis ovatis integerrimis tomentosis, stolonibus repentibus, scapo unifloro. *Lin. Syst. Vegetab.* p. 597. *Sp. Pl.* p. 1125. *Fl. Suec.* n. 697.

HIERACIUM caule unifloro, stolonibus reptantibus, foliis petiolatis, ovatis, longe pilosis, subtus tomentosis. *Haller. Hist.* n. 53.

HIERACIUM *Pilosella. Scopoli Fl. Carniol.* n. 966.

PILOSELLA major repens hirsuta. *Bauh. Pin.* 262.

PILOSELLA minor vulgaris repens. *Parkinson*, 690.

PILOSELLA repens. *Ger. emac.* 638. *Raii Syn.* p. 170. Common creeping Mouse-ear. *Hudson Fl. Angl.* p. 343. *Lightfoot Fl. Scot.* p. 436.

RADIX perennis, fibrosa, fibris pallide fuscis.

ROOT perennial, fibrous, the fibres of a pale brown colour.

FOLIA petiolata, ovata, sæpe oblonga, integerrima, superne virentia, scabriuscula, inferne alba, tomentosa, utrinque pilis longis obsita, e centro foliorum ad basin caulis, oriuntur stolones teretes, repentes, hirsuti, foliosi, foliis utplurimum lanceolatis.

LEAVES standing on foot-stalks, ovate, often oblong, perfectly entire, on the upper side green and roughish, on the under side white and downy, on both sides beset with long hairs, from the center of the leaves, at the base of the stalk, spring one or more round, creeping, hirsute, leafy runners with leaves for the most part lanceolate.

SCAPUS: ex una planta seu rosula exsurgit scapus plerumque solitarius, palmaris, spithamæus et ultra, erectus, teres, nudus, fistulosus, inferne pilosus, superne hirsutus, uniflorus.

STALK: from a single plant or off-set arises generally one solitary stalk, from four to seven inches or more in height, upright, round, naked, hollow, below hairy, above hirsute, supporting a single blossom.

FLORES pallide lutei, seu sulphurei, extimis flosculis inferne purpurascentibus.

FLOWERS of a pale yellow or sulphur colour, the outermost florets purplish on the under side.

CALYX communis imbricatus, squamis pluribus, linearibus, valde inæqualibus. *fig.* 1, 2. hirsutis, pilis nigricantibus, ad lentem globiferis.

CALYX: the common Calyx imbricated, the scales numerous, linear, very unequal. *fig.* 1, 2. hirsute, the hairs blackish, and when magnified globular at the extremity.

COROLLA composita, imbricata, uniformis; *Corollulis* hermaphroditis, numerosis, æqualibus; *proprio* monopetala, *Tubus* pappo longior, albus, lanatus. *Limbus* planus, quinquedentatus longitudine tubi. *fig.* 3, 4.

COROLLA compound, imbricated and uniform, the *Florets* hermaphrodite, numerous, equal, and monopetalous. The *Tube* longer than the pappus, white and woolly. The *Limb* flat, having five teeth, the length of the tube. *fig.* 3, 4.

STAMINA: FILAMENTA quinque capillaria, brevissima. ANTHERÆ in tubum cylindricum, flavum coalitæ. *fig.* 5.

STAMINA: five capillary FILAMENTS, very short. ANTHERÆ united in a cylindrical, yellow tube. *fig.* 5.

PISTILLUM: GERMEN oblongum: STYLUS filiformis, longitudine staminum. STIGMATA duo, revoluta. *fig.* 6.

PISTILLUM: GERMEN oblong. STYLE filiform, the length of the stamina. STIGMATA two, rolled back. *fig.* 6.

SEMEN oblongum, nigricans, striatum. *Pappus* semine duplo longior, sessilis, simplex, *fig.* 7, 8.

SEED oblong, blackish, striated. *Down* twice the length of the seed, sessile, and simple. *fig.* 7, 8.

The ancient botanists fancied some similitude betwixt the hairy leaves of this plant and a mouse's ear, whence its name.

Of the whole genus of Hawk-weed this is the most universally common in this country. It delights in dry and exposed situations, which are frequently rendered barren by long continued drought, the sunny bank, the lifeless wall, and arid heath, are often enamelled with its lively flowers, which appear in *May* and *June*, and frequently produce a second crop late in the summer.

It has been received into the shops under the name of *Auricula muris*, and considered as possessing an astringent quality; but at present, in this respect, is but little regarded.

SIMON PAULI discovered on its roots small tubercles, which he considered as the eggs of an unknown insect; these have since proved to be a *Coccus* (*Coccus Pilosella, Lin.*) nearly related to the *Coccus polonicus*, an insect used in dying. We have not heard of its having been observed here.

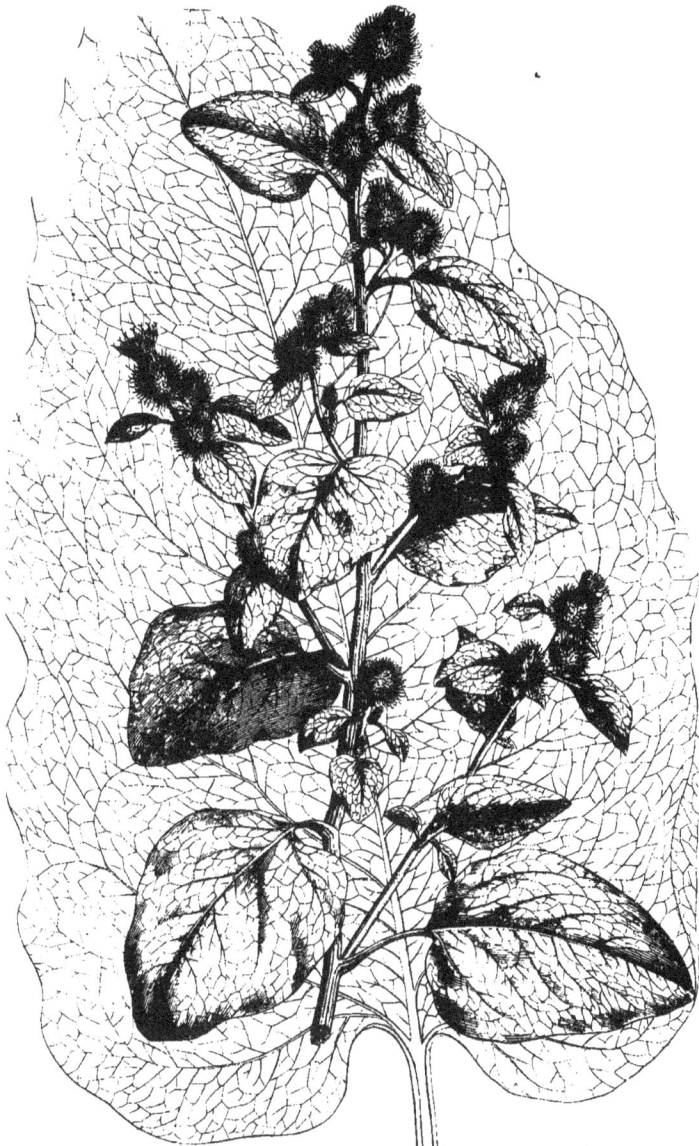

(Arctium Lappa.)

ARCTIUM *Lin. Gen. Pl.* SYNGENESIA POLYGAMIA ÆQUALIS.

Cal. globofus; fquamis apice hamis inflexis.

Raii Syn. Gen. 9. HERBÆ FLORE EX FLORIBUS FISTULARIBUS COMPOSITO SIVE CAPITATÆ

ARCTIUM *Lappa* foliis cordatis inermibus petiolatis. *Lin. Syft. Vegetab.* p. 603. *Sp. Pl.* 1143. *Pl. Suec.* n. 712.

LAPPA *Haller Hift.* n. 161.

LAPPA major feu Arctium Diofcoridis. *Bauh. Pin.* 198.

PERSONATA five Lappa major aut Bardana. *J. B.* III. 570.

BARDANA major. *Ger. emac.* 809.

BARDANA vulgaris major. *Park.* 1222. *Raii Syn.* 197. Great Burdock, Clot-burr. *Hudfon. Fl. Angl. ed. 2.* p. 348. Lightfoot *Fl. Scot.* p. 197. *Order Fl. Dan.* t. 642.

RADIX biennis, fimplex, profunde in termam defcendens, foris nigricans, intus alba.

ROOT biennial, fimple, penetrating deeply into the earth, externally blackifh, internally white.

CAULIS erectus, tripedalis et ultra, craffitudine pollicis, ad bafin ufque ramofiffimus, teres, ftristo-fulcatus, villofus, purpurafcens.

STALK upright, three feet high and more, the thicknefs of one's thumb, branched quite down to the bottom, round, grooved, but not deeply, hoary and purplifh.

FOLIA ampliffima, petiolata, alterna, cordata, fubtus albida, venofa, margine fubundulatâ, crenulatâ.

LEAVES very large, ftanding on footftalks, alternate, heart-fhaped, whitifh underneath, veiny, the edge fomewhat waved and notched.

PETIOLI foliis breviores, fulcato-angulati, villofi.

LEAF-STALKS fhorter than the leaves, angular or grooved, and hoary.

FLORES purpurei, pedunculati, erecti, ramulis alterne difpofiti, fuperne in capitula laxa collecti.

FLOWERS purple, ftanding on footftalks, upright, difpofed alternately on the branches, and at the tops of them collected into loofe heads.

CALYX communis globofus, imbricatus, glaber, filis araneofis intertextus, fquamis exterioribus apice hamatis, hamis nitidis, acutiffimis, fubinflexis, *fig.* 1, 2. Interioribus linearibus, apice vivide purpureis, fimplicibus, fine hamis.

CALYX common to all the florets globular, imbricated, furface fhining, cobwebby, the exterior fcales hooked at the extremity, hooks fhining, very fharp, and fomewhat bent in, *fig.* 1, 2. Inner fcales linear, tops of a bright purple colour, fimple, without hooks.

COROLLA compofita, calyce longior, tubulata, *propria* infundibuliformis, tubo filiformi, albo, limbo tubulofo-campanulato, purpureo, quinque-fido, acuto, erecto.

COROLLA compound, longer than the calyx, tubular, *Florets* funnel-fhaped, tube filiform, white, limb tubular, and fomewhat bell-fhaped, of a purple colour, divided into five upright, pointed fegments.

STAMINA: FILAMENTA quinque alba, capillaria; ANTHERÆ cærulefcentes, aut violaceæ, in tubum extra corollam coalitæ.

STAMINA: five white capillary FILAMENTS; ANTHERÆ blueifh, or violet coloured, united into a tube, which projects beyond the corolla.

PISTILLUM: GERMEN fubtriquetrum, album, incurvatum; STYLUS albus, ftaminibus longior, utrinque fulcatum; STIGMA bifidum.

PISTILLUM: GERMEN fomewhat three-cornered, white, bending inward; STYLE white, longer than the ftamina, grooved on each fide; STIGMA bifid.

SEMINA oblonga, bafi attenuata, punctis prominentibus coronata, compreffa, fubangulata, reticulatorugofa, exterioribus incurvatis. *Pappus* obfoletus, rigidulus. *fig.* 3.

SEEDS oblong, tapering towards the bafe, crowned with fmall prominent points, flattened, fomewhat angular, furface wrinkly, in the form of net-work, outermoft feeds bending inward. *Down* obfolete, fomewhat rigid. *fig.* 3.

The antient botanifts divided the Burdock, which they diftinguifhed by the feveral names of *Lappa* * *Perfonata*, and *Arctium* or *Arcium* into two principal fpecies, viz. the *Lappa major Arcium Diofc. C. B.*; and the *Lappa major montana capituli tomentofis five Arctium Diofc. C. B.*; both of which are admirably figured by MATTHIOLUS on wood, as indeed are moft of the medicinal plants; later botanifts have made the fpecies much more numerous; in the third edition of RAY's *Synopfis* we find no lefs than fix fpecies and one variety, viz.

1. *Lappa major capitulo glabro maximo.*
2. *Lappa major Arcium Diofcoridis.*
3. *Lappa major capituli parvi glabri.*
4. *Lappa major montana capitulis tomentofis; five Arctium Diofcoridis.*

* *Lappa* dici poteft vel ἀπὸ τῦ λαβεῖν prehendere vel ἀπὸ τῦ λάπτω, i. e. lambere, quod prætereuntium veftibus adhæreat. *Perfonata* autem dicitur, quod folia ejus prægrandia veluti larvæ aut perfonæ vice obtendi folita effent. Veteribus *Arction* aut *Arcion* dicta credstur, verum ratio nominis ignoratur. *Raii Hift.* p. 332.

5. *Lappa*

5. *Lappa major montana, capitulis minoribus, rotundioribus & magis tomentofis.*
6. *Lappa major ex omni parte minor, capitulis parvis eleganter reticulatis.*
Lappa vulgaris major capitulis foliofis. Var.
All thefe are however confidered by the botanifts of the prefent day as one fpecies, to which may be added a variety with white flowers, which often occurs.

The Burdock is a very common plant by way fides, and in wafte places, and flowers in July and Auguft.

In the fize of its leaves it may fometimes difpute the palm with the Butterburr; painters often introduce them in the foregrounds of their pictures, which they are admirably well calculated to embellifh.

No fort of cattle relifh the foliage of this plant, but fnails, flugs, and many fpecies of caterpillars feed on it delicioufly; the pith produces its particular moth, which does not appear to be defcribed by LINNÆUS; but is moft accurately figured by SEPP*, and called by Aurelians the *Mottled Orange*, the caterpillar of this moth changes into chryfalis about the beginning of Auguft, during which month it may be found in that ftate by fplitting the ftalks of fuch plants as appear ftinted in their growth; the moth comes out about the end of Auguft, and is one of thofe whofe bodies are extremely apt to become greafy, to prevent which the body fhould be carefully opened on the under fide, and its contents taken out previous to its being placed in the cabinet. A fmall larva, not peculiar to this plant, feeds alfo betwixt the coats of the leaf.

The feeds, like thofe of the thiftle tribe, are fought for by feveral of the feathered fongfters, and are even recommended to fatten poultry †.

The microfcope informs us, *vide fig.* 1, 2. why the burrs adhere fo clofely to one's cloaths, and why boys, who divert themfelves by throwing them at one another, have fo much difficulty in extricating them from their hair.

As a weed it is not fo formidable as it appears to be, being a biennial the hufbandman has only to deftroy its feedlings.

The root and ftalks are efculent and nutritive; the ftalks for this purpofe fhould be cut before the plant flowers, the rind peeled off, and then boiled and ferved up in the manner of Chardoons, or eaten raw, as a fallad, with oil and vinegar. *Lightfoot Fl. Scot.* p. 446.

The feeds have a bitterifh, fubacrid tafte, they are recommended as very efficacious diuretics, given either in the form of emulfion or in powder to the quantity of a drachm. The roots tafte fweetifh, with a flight aufterity and bitterifhnefs; they are efteemed aperient, diuretic and fudorific, and faid to act without irritation, fo as to be fafely ventured on in acute diforders. Decoctions of them have of late been ufed in rheumatic, gouty, and other diforders, and preferred by fome to thofe of Sarfaparilla. LEWIS's *Difp.* p. 101.

* *Nederlandfche Infecten.* t. 3. † *Stirp. indig. dragm.* p. 113.

CICHORIUM INTYBUS. BLUE SUCCORY.

CICHORIUM *Lin. Gen. Pl.* SYNGENESIA POLYGAMIA ÆQUALIS.
    Recept. fubpaleaceum. *Cal.* calyculatus. *Pappus* fub-5-dentatus, obfolete
    pilofus.

    Raii Syn. Gen. 6. HERBÆ FLORE COMPOSITO NATURA PLENO LACTESCENTES.
CICHORIUM *Intybus* floribus geminis feffilibus, foliis runcinatis. *Lin. Syft. Vegetab.* p. 601. *Sp. Pl.*
    p. 1142. *Fl. Suec.* n. 711.
CICHORIUM foliis pinnatis, pinnis triangularibus dentatis, floribus feffilibus. *Haller Hift.* 1.
CICHORIUM *Intybus. Scopoli Fl. Carn.* n. 991.
CICHORIUM fylveftre five officinarum, *Bauhin Pin.* 126.
INTYBUS fylveftris. *Camer. epit.* 285.
CICHORIUM fylveftre. *Ger. emac.* 284. *Parkinf.* 776. *Raii Syn.* p. 172. Wild Succory. *Hudfon*
    Fl. Angl. ed. 2. p. 348.

RADIX perennis, externe lutefcens, fufiformi-cylindrica, etiam ramofa, craffitie digiti, fpithamæa, fæpe pedalis, defcendens, fibrillofa, fibrillis fparfis, lactefcens, lacte albo.

ROOT perennial, externally of a yellowifh colour, tapering very gradually to a point, alfo branched, the thicknefs of the finger, a fpan, and often a foot in length, ftriking downward, furnifhed with few fmall fibres, milky, the milk of a white colour.

CAULIS pedalis, ad tripedalem, erectus, rigidus, tortuofus, fcabridus, plerumque ramofiffimus.

STALK from one to three feet in height, upright, rigid, crooked, roughifh to the touch, and generally very much branched.

FOLIA radicalia plurima, taraxaci, fubafpera, caulina læviora, fubamplexicaulia, alterna.

LEAVES at the root numerous, like thofe of Dandelion, roughifh, thofe of the ftalk fmoother, alternate, partly furrounding the ftem.

FLORES plerumque bini, fpeciofi, feffiles, e foliorum fupremorum alis.

FLOWERS growing generally in pairs, fhowy, feffile, fpringing from the alæ of the uppermoft leaves.

CALYX *communis* calyculatus, fquamis exterioribus quinque, ovatis, acutis, fubpatentibus, pilis glanduliferis ciliatis; interioribus octo circiter, lineari-lanceolatis, æqualibus, cylindrum angulofum, vifcofum, conftituentibus. *fig.* 1, 2, 3, 4.

CALYX common to many florets, compof'd of a double fet of fquamæ or leaves, the outermoft of which are five in number, ovate, pointed, fomewhat fpreading, edged with glandular hairs, the innermoft about eight, narrow, equal, forming an angular, clammy cylinder. *fig.* 1, 2, 3, 4.

COROLLA compofita, plana, uniformis. Corollulis hermaphroditis, viginti circiter, cæruleis, *Tubus* cylindricus, brevis, albus, apice dilatatus et pilofus; *Limbus* planus, quinque-dentatus, fubtus nervofus et villofus. *fig.* 5.

COROLLA compound, flat, regular, Florets hermaphrodite, about twenty in number, of a blue colour; *Tube* cylindrical, fhort, white, dilated at top and hairy; *Limb* flat, with five teeth at the extremity, on the under fide rib'd and villous. *fig.* 5.

STAMINA: FILAMENTA quinque alba, capillaria, manifefte libera; ANTHERÆ faturate cæruleæ, in tubum cylindricum, angulatum coalitæ. *fig.* 6.

STAMINA: five FILAMENTS, of a white colour, very flender, manifeftly unconnected; ANTHERÆ of a deep blue colour, forming an angular, cylindrical tube. *fig.* 6.

PISTILLUM: GERMEN fubconicum, album, pilis breviffimis coronatum; STYLUS filiformis, albus; STIGMATA duo, cærulea, revoluta. *fig.* 7.

PISTILLUM: GERMEN fomewhat conic, crowned with very fhort hairs; STYLE thread-fhaped, white; STIGMATA two, of a blue colour, and rolled back.

SEMINA plurima, in fundo calycis, nuda, fubpentagona, margine pilis breviffimis ciliata. *fig.* 8, auct.

SEEDS numerous in the bottom of the calyx, naked, irregularly five cornered, the edge crowned with very fhort hairs. *fig.* 8. magnified.

That beautiful plants are often noxious weeds, agriculturally confidered, we have already noticed in the Biftort, the field Convolvulus, the corn Poppy, and the perennial Perficaria; the blue Succory adds another to the catalogue.

Batterfea Fields, which exhibit bad hufbandry in perfection, produce this plant moft plentifully; it flowers in *July, Auguft* and *September*; like the docks it increafes itfelf much by feed, and is to be extirpated in the fame manner.

Some botanifts have erroneoufly fuppofed this fpecies of Succory to be the Endive in its wild ftate, but its ftrong perennial root fufficiently evinces the contrary. The *Cichorium Endivia,* which is an annual or biennial, and grows wild in the Corn-fields of *Spain,* together with the *Intybus**, is undoubtedly the parent of the cultivated *Endive,* it is not fo clear which of the two is the plant celebrated by HORACE as conftituting a part of his fimple diet,

    —— me pafcunt Olivæ
    Me Cichorea, levefque Malvæ.

It is not unfrequently found wild with white flowers, and it has been difcovered that the fine blue colour of the petals is convertible into a brilliant red by the acid of Ants†; Mr. MILLER the Engraver affured me, that in *Germany* the boys often amufed themfelves in producing this change of colour by placing the bloffoms in an ant hill.

Wild Succory is an ufeful detergent, aperient, and attenuating medicine; acting without much irritation, tending rather to cool than heat the body, and at the fame time corroborating the tone of the inteftines. The juice taken in large quantities fo as to keep up a diarrhœa, and continued for fome weeks, has been found to produce excellent effects in fcorbutic and other chronical diforders. *Lewis's Difp.* p. 125.

---

    * *D'Affi Stirp. Arrogan.* p. 113.   † *Trag. ad Brunfels.* II. p. 274.

Cichorium Intybus.

BIDENS *Lin. Gen. Pl.* Syngenesia Polygamia Æqualis.

Recept. paleaceum. *Pappus* ariſtis erectis ſcabris. *Cal.* imbricatus. *Cor.* rarius floſculo uno alterne radiante inſtruitur.

Raii Syn. Gen. 8. Herbæ flore composito discoide seminibus pappo destitutis corymbiferæ dictæ.

BIDENS *tripartita* foliis trifidis, calycibus ſubfoliofis ſeminibus erectis. *Lin. Syſt. Vegetab.* p. 610. *Sp. Pl.* 1165. *Fl. Suec.* 283. *Lappon.* p. 234.

BIDENS foliis petiolatis trilobatis et quinque lobatis ſerratis floribus circumvallatis. *Haller Hiſt.* n. 121.

BIDENS *tripartita. Scop. Fl. Carn.* n. 1090.

VERBESINA ſeu Cannabina aquatica flore minus pulchro, elatior et magis frequens. *J. B.* II. 1073.

CANNABINA aquatica folio tripartito diviſo. *Bauh. pin.* 321.

EUPATORIUM cannabinum fœmina, *Ger. emac.* 711.

EUPATORIUM aquaticum duorum generum. *Parkinſ.* p. 595. *Raii Syn.* p. 187. Water Hemp-Agrimony, with a divided Leaf. *Hudſon. Fl. Angl. ed.* 2. p. 355. *Lightfoot Fl. Scot.* p. 461.

RADIX annua, ſimplex, fibroſa, fibris albidis.

CAULIS pedalis ad tripedalem, erectus, ramoſus (ramis oppoſitis), teretiuſculus, modice ſolcatus, rubens, ſolidus, glaber, ſcabriuſculus.

FOLIA oppoſita, petiolata, connata, glabra, tripartita, aut etiam quinque partita, laciniis profunde ferratis, ſuprema indiviſa, dentato-ſerrata, aut etiam integra, pilis haud infrequenter ciliata.

FLORES lutei, terminales, ſubnutantes.

CALYX : Foliola plura, plerumque integra, lanceolata, ciliata, flores involucri inſtar ambientia ; ſquamæ calycis communis ovato lanceolatæ, integræ, lineis plurimis, nigricantibus, parallelis, pictæ, marginibus flaveſcentibus. *fig.* 1.

COROLLULÆ hermaphroditæ, tubuloſæ, infundibuliformes, luteæ, ſtriis quinque purpureis externe notatæ ; limbo quinquefido, ſuberecto. *fig.* 2.

STAMINA : Filamenta quinque capillaria ; Antheræ in tubum cylindricum coalitæ. *fig.* 3.

PISTILLUM : Germen ſubcompreſſum, angulatum, ſuperne latius, ariſtis tribus plerumque inſtructum, unica breviore ; *fig.* 3. Stylus ſimplex, longitudine ſtaminum ; Stigmata duo oblonga, reflexa. *fig.* 4, 5.

SEMEN oblongum, compreſſum, angulatum, fuſcum, ariſtis duabus ſeu tribus retrorſum ſcabro hamatis inſtructum. *fig.* 6.

RECEPTACULUM paleaceum, planum, paleis lanceolato-linearibus, lineatis, deciduis. *fig.* 7.

ROOT annual, ſimple and fibrous, fibres whitiſh.

STALK from one to three feet high, upright, branched, (the branches oppoſite), roundiſh, moderately grooved, of a reddiſh colour, ſolid, ſmooth to appearance, but ſlightly rough to the touch.

LEAVES oppoſite, ſtanding on footſtalks, which unite at the baſe, ſmooth, divided into three, and ſometimes five ſegments, which are deeply ſerrated, the uppermoſt leaves undivided, either indented at the edge, or entire, and not unfrequently edged with hairs.

FLOWERS yellow, terminal, drooping a little.

CALYX : ſeveral, ſmall, lanceolate leaves, generally entire, but edged with hairs ſurrounding the flowers like an involucrum ; the ſcales of the calyx common to all the florets are ovate and pointed, entire at the edge, and painted with numerous blackiſh lines, the edges are yellowiſh. *fig.* 1.

FLORETS hermaphrodite, tubular, funnel-ſhaped, of a yellow colour, marked externally with three purpliſh ſtripes, the limb divided into five ſegments, which are nearly upright. *fig.* 2.

STAMINA : five capillary Filaments ; Antheræ united into a cylindrical tube. *fig.* 3.

PISTILLUM : Germen flattiſh, angular, broadeſt at top, generally furniſhed with three awns, of which one is ſhorter than the reſt ; *fig.* 3. Style ſimple, the length of the ſtamina ; Stigmata two, oblong, turning back. *fig.* 4, 5.

SEED oblong, flat, angular, brown, furniſhed with two or three awns, which are hooked or barbed downward. *fig.* 6.

RECEPTACLE chaffy and flat, ſcales or chaff, narrow, marked with lines and deciduous. *fig.* 7.

This ſpecies of *Bidens* is much more common than the *cernua*, as that is generally found in the water, this more frequently occurs on the borders of ponds, rivulets, &c. where it flowers in the months of Auguſt and September, at the cloſe of which it ripens its ſeeds.

It is obviouſly diſtinguiſhed from the *cernua* by having its leaves, for the moſt part, divided into three ſegments, whence its name ; this character is more to be depended on than the uprightneſs of its flowers, as they generally droop a little when the plant is in perfection.

Linnæus, and other writers, recommend it as a plant that will dye both linen and woollen of a yellow colour, for this purpoſe the yarn or flax muſt be firſt ſteeped in allum-water, then dried and ſteeped in a decoction of the plant, and afterwards boiled in the decoction. *Haller. Hiſt. Helv.* p. 52.

Bidens tripartita.

jasione montana.

JASIONE. *Lin. Gen. Pl.* Syngenesia Monogamia.

> *Cal.* communis 10-phyllus. *Cor.* 5-petala, regularis. *Capf.* infera, bilocularis.

JASIONE *montana. Lin. Syst. Vegetab.* p. 666. *Spec. Pl.* p. 1317. *Fl. Suec.* n. 782.

RAPUNCULUS foliis linearibus subasperis, spica planiuscula, petalis liberis. *Haller Hist.* n; 678;

RAPUNCULUS scabiosæ capitulo cæruleo. *Bauhin Pin.* 92.

RAPUNTIUM montanum capitatum leptophyllon. *Col. Ecphr.* 1. p. 226: t. 227.

SCABIOSA globularis quam ovinam vocant *J. B.* III. 12.

SCABIOSA minima hirsuta. *Ger. emac.* 723. *Raii Syn.* p. 278. Hairy Sheep's Scabious, or rather Rampions with Scabious Heads. *Hudson Fl. Angl. ed.* 2. p. 377. *Lightfoot Fl. Scot.* p. 377.

RADIX annua, lignosa, albida, fibrosa.

CAULES plures, suberecti, spithamæi, etiam pedales et ultra, rigiduli, ramosi, hirsuti.

FOLIA plurima, sessilia, lineari-lanceolata, obtusiuscula, undulata, hirsuta.

FLORES capitati, cærulei, summitatibus ramorum insidentes.

CALYX : *Perianthium commune* polyphyllum ; foliolis alternis, interioribus angustioribus, includens flores plurimos pedunculis brevissimis adnexos, persistens. *fig.* 1.

> *Perianthium proprium* quinquefidum, superum, persistens.

COROLLA propria pentapetala : *Petalis* lanceolatis, erectis, basi connexis. *fig.* 2.

STAMINA quinque. FILAMENTA subulata, brevia. ANTHERÆ quinque, oblongæ, basi connexæ. *fig.* 3.

PISTILLUM : GERMEN subrotundum, inferum. STYLUS filiformis, longitudine Corollæ. STIGMA clavatum, purpureum. *fig.* 4, 5.

PERICARPIUM : CAPSULA subrotunda, quinquangularis, coronata calyce proprio, bilocularis.

SEMINA plura, subovata.

ROOT annual, rigid, whitish and fibrous.

STALKS several, nearly upright, about a span in length, but sometimes a foot or more, rather rigid, branched, and beset with short rough hairs.

LEAVES numerous, sessile, between linear and lanceolate, bluntish, waved and hirsute.

FLOWERS of a blue colour, growing in little heads on the tops of the branches.

CALYX : the *Perianthium common to all the florets* composed of many leaves, which are alternate, those of the inner-row narrowest, including numerous flowers sitting on very short footstalks, and permanent. *fig.* 1.

> *The Perianthium of each floret* deeply divided into five segments above the germen, and permanent.

COROLLA : each floret composed of five lanceolate, upright *Petals*, connected at the base. *fig.* 2.

STAMINA : five tapering short FILAMENTS ; ANTHERÆ five, oblong, connected at the base. *fig.* 3.

PISTILLUM : GERMEN roundish, below the Corolla. STYLE filiform, the length of the corolla. STIGMA club-shaped and purplish. *fig.* 4, 5.

SEED-VESSEL : a roundish CAPSULE, having five angles with two cavities, and crowned by the calyx proper to it.

SEEDS numerous, somewhat ovate:

This little plant, which in its general appearance so much resembles a Scabious, is very common on dry, sandy ground, especially about *Coomb Wood*, and *Hampstead*, and most hilly situations near *London*, and elsewhere.

It varies much in size, and is sometimes, though very rarely, found with white blossoms.

It flowers from *June* to *August*.

LINNÆUS remarks, that Bees are particularly fond of its flowers.

OPHRYS SPIRALIS. LADIES TRACES.

OPHRYS *Lin. Gen. Pl.* GYNANDRIA DIANDRIA.

Nectarium fubtus fubcarinatum.

Raii Syn. Gen. 21. HERBÆ RADICE BULBOSA PRÆDITÆ.

OPHRYS *fpiralis* bulbis aggregatis oblongis, caule fubfoliofo, floribus fecundis, nectarii labio indivifo crenato. *Lin. Syft. Vegetab.* p. 677. *Sp. Pl.* 1340.

EPIPACTIS bulbis cylindricis, fpica fpirali, labello crenulato. *Haller. Hift.* n. 1294.

SERAPIAS *fpiralis. Scopoli Flor. Carn.* n. 1125.

ORCHIS fpiralis alba odorata. *J. B.* II. 769.

TRIORCHIS alba odorata minor, atque etiam major. *Bauhin. Pin.* 84.

TRIORCHIS. *Ger. emac.* 218. *Parkins.* 1354 *Raii Syn.* p. 378. Triple Ladies Traces. *Hudfon Fl. Angl.* p. 388.

RADIX fit uno, duobus, tribus, quatuorve bulbis, oblongis, acuminatis, villofis.

FOLIA radicalia quatuor, et ultra, fupra terram expanfa, ovata, acuta, hinc convexa, inde cava, femuncinm leta, ad lentem punctata, obfolete nervofa.

SCAPUS fpithamæus, foliofus, foliis vaginantibus, pubefcentibus, margine membranaceis.

FLORES ex albo-virefcentibus, odorati, quindecim et ultra, *fpiræ modo difpofiti.*

BRACTÆA oblonga, acuminata, cava, villofa, germinis cum dimidio floris longitudine. *fig.* 1.

PETALA quinque, alba, fubæqualia, villofula, tria fuperiora fubcoadunata, recta, duo lateralia carinata, lanceolata, *fig.* 2, 3.; labellum *Nectarii* obtufum, crenulatum, intus viridulum, concavum, *fig.* 4. auct. *fig.* 5.

GERMEN feffile, ovatum, lineis duabus lateralibus extantibus notatum. *fig.* 6.

ROOT confifts of one, two, three, or four oblong, pointed, villous bulbs.

LEAVES next the root four, or more, fpread out on the ground, ovate, pointed, convex on one fide, and concave on the other, half an inch in breadth, dotted when magnified, and faintly ribbed.

STALK fix or feven inches high, leafy, leaves fheathy, downy, and membranous at the edge.

FLOWERS of a greenifh white colour, fragrant, fifteen and more in number, *fpirally difpofed.*

FLORAL-LEAF oblong, pointed, hollow, villous, of the length of the germen, and half the flower, *fig.* 1.

PETALS five, white, nearly equal, fomewhat villous, the three uppermoft very flightly connected together, ftraight, the two fide ones keeled and lanceolate, *fig.* 2, 3.; the lip of the *Nectary* blunt, finely notched, green within and hollow, *fig.* 4. magnified, *fig.* 5.

GERMEN feffile, ovate, marked with two protuberant fide lines. *fig.* 6.

The Rev. Dr. GOODENOUGH, of *Ealing,* kindly communicated to us this plant, having found it fparingly on *Hanwel Heath,* near *Ealing*: though fcarce with us, in many parts of *England,* efpecially the more northern, it is not uncommon. It grows in paftures, both dry and moift, and does not particularly affect a chalky foil. In the garden it grows more readily than moft of its tribe, and flowers later, its ufual month of blowing being *September.*

The protuberant germina, placed regularly one above another, fomewhat refemble plaited hair, whence, perhaps, its name of *Ladies Traces.* The flowers are fragrant, and, by the fpiral manner in which they grow, form a curious fpecific character.

Baron HALLER, who has taken infinite pains with the plants of this tribe, has not very happily exprefed this fpecies; his artift appears to have had an unnatural fpecimen to copy from.

The Ladies Traces varies much in fize as well as in the number of its roots.

Ophrys spiralis.

Carex riparia.

CAREX RIPARIA.   GREAT OR COMMON CAREX.

CAREX *Lin. Gen. Pl.* MONOECIA TRIANDRIA.

    MASC. Amentum imbricatum. *Cal.* 1. phyllus. *Cor.* o.

    FEM. Amentum imbricatum. *Cal.* 1. phyllus. *Cor.* o. *Nectarium* inflatum, 3. dentatum. *Stigm.* 3.

    SEM. Triquetrum, intra nectarium.

Raii Synop. Gen. 28. HERBÆ GRAMINIFOLIÆ NON CULMIFERÆ FLORE IMPERFECTO SEU STA-MINEO.

CAREX *riparia* spicis masculis pluribus triquetris nigricantibus, acutis, squamis aristato acuminatis, capsulis subinflatis, bicornibus.

CAREX *acuta* spicis masculis pluribus, femineis subpedunculatis, erectis, capsulis ovato-lanceolatis aristato-acuminatis furcatis. *Hudson Fl. Angl. p.* 413.

CAREX spicis masculis ternis, femineis numerosis, erectis, brevissime petiolatis, capsulis bicornibus. *Haller, hist. n.* 1404. et forsan 1398 et 1399.

CAREX *acuta. Lightfoot, Fl. Scot. p.* 565.

GRAMEN cyperoides cum paniculis nigris. *J. B.* 2. 494. *Raii Hist.* 1292.

GRAMEN cyperoides latifolium spica rufa sive caule triangulo. *Bauh. Pin.* 6.

GRAMEN cyperoides. *Ger. emac.* 12.

GRAMEN cyperoides majus latifolium. *Park.* 1265.

Raii Syn. 417. Great vernal Cyperus-grass.

CYPEROIDES aquaticum, maximum, foliis vix unciam latis, caule exquisite triangulari, spicis habitioribus, erectis, squamis in aristam longius productis, capsulis oblongis, bifidis. *Michel. Nov. Gen. Tab.* 31. *fig.* 7. et 6.

RADIX perennis, repens.	‡ROOT perennial and creeping.
CULMUS in aquosis bi seu tripedalis, foliosus, nodosus, strictus, triqueter, angulis acutis, asperis.	‡STALK in wet situations two or three feet high, leafy, jointed, striated, the angles sharp and rough.
FOLIA semunciam lata, glauca, carinata, ad margines carinamque aspera, vaginantia, vagina una cum inferiore parte folii pulchre reticulata.	‡LEAVES half an inch broad, glaucous, keeled, the keel as well as the edges rough, sheathing the stalk, the sheath, together with the lower part of the leaf, beautifully reticulated.
SPICÆ masculæ et femineæ distinctæ, *masculæ*, plerumque, tres, ad quinque, erectæ, nigricantes, triquetræ, acutæ, congestæ, bracteatæ, suprema biunciali, inferioribus brevioribus inæqualibus, *femineæ* tot quot masculæ, ovato-acutæ, pedunculatæ, plerumque erectæ, aliquando etiam pendulæ, supremis sessilibus, androgynis.	‡SPIKES of the male and female distinct, those of the *male* generally from three to five, upright, blackish, three-cornered, pointed, clustered and furnished with floral leaves, the uppermost about two inches in length, the lowermost shorter and unequal; *female* spikes as numerous as those of the *male*, ovate, pointed, standing on footstalks, generally upright, but sometimes pendulous, the uppermost sessile and androgynous.
FLOS MASC.	MALE FLOWER.
CALYX: *Squamæ* plurimæ, imbricatæ, lanceolatæ, aristato-acuminatæ, e nigro purpurascentes. *fig.* 1.	CALYX: *Scales* numerous, imbricated, lanceolate, running out to a long beard-like point, of a purplish black colour. *fig.* 1.
STAMINA: FILAMENTA tria, filiformia, alba; AN-THERÆ tenues, luteæ, mucronatæ. *fig.* 2.	STAMINA: three FILAMENTS, thread-shaped and white; ANTHERÆ slender, yellow, and terminated by a short point. *fig.* 2.
FLOS FEM.	FEM. FLOWER.
CALYX: *Squamæ* ut in masc. inferne vero latiores et superne magis luculenter aristatæ, aristâ serrulatâ. *fig.* 3.	CALYX: *Scales* as in the male, but broader below, and more evidently bearded above, the awn finely sawed or toothed. *fig.* 3.
NECTARIUM germen continens, ovatum, glabrum, bicorne. *fig.* 4. auct. demum inflatum, acuminatum, striatum, fuscum. *fig.* 6. magn. natur.	NECTARY containing the germen, ovate, smooth, with two horns, *fig.* 4. magn. finally inflated, pointed, striated, and of a brown colour. *fig.* 6. nat. size.
PISTILLUM: Germen parvum, ovatum, glabrum; STYLUS filiformis, nudus, nectario paulo longior; STIGMATA tria, villosa, alba, subulata. *fig.* 5. auct.	PISTILLUM: GERMEN small, ovate, smooth; STYLE filiform, naked, a little longer than the nectary; STIGMATA three, villous, white and tapering. *fig.* 5 magnif.
SEMEN unicum, triquetrum, intra Nectarium. *fig.* 7. mag. nat.	SEED single, three-cornered, inclosed in the Nectary, *fig.* 7. nat. size.

In a former number of this work we gave a figure and description of the *Carex pendula*, one of the largest, as well as most distinct species of this genus; we here present our readers with three more of this numerous and difficult tribe. Our motive for publishing them in the same number is, that they may the more readily be compared together, and their several distinguishing characters be more forcibly impressed.

In herborizing it is a practice with me to endeavour at acquiring a perfect knowledge of every plant which occurs in all its possible varieties; the greater the difficulty I find in the attempt, the more minute is my enquiry. These investigations have to my great satisfaction often terminated in some new discovery, which has placed the plant in a more conspicuous light than before; such has been the happy result in the present instance. In passing through *Battersea* meadows I had frequently noticed the three Carices here figured, which I was taught to consider as the same species, varying only from particular circumstances, but so great was the variation, that I never could perfectly reconcile myself to the idea. I shall here relate the several characters which struck me first, and gave me the idea of their being different. It was the pointed, triangular, black heads or male spikes of the *riparia*, the bluntness not only of

the

b

CAREX *acuta* fpicis mafculis pluribus, obtufis, fquamis obtufiufculis, caule acutangulo.
CAREX *acuta* fpicis mafculis pluribus, femineis fubfeffilibus, capfulis obtufiufculis. *Lin. Syft. Vegetab.*
p. 706. *Sp. i l.* p. 1388 *Fl. Suec.* n. 857.
CAREX *glaura Scopoli Fl. Carn.* n. 1157 ?
CYPEROIDES foliis Caryophylleis, caule exquifite triangulari, fpicis habitioribus, fquamis curtis, obtufé mucronatis, capfulis turbinatis, brevibus, confertis. *Michel Nov. Gen.* p. 62. *tab.* 32. *f.* 11.
GRAMEN cyperoides foliis caryophylleis vulgatiffimum. *Raii bift.* 1292.
CAREX cæfpitofa var β. *Ligbijooi Fl. Scot.* ?

RADIX perennis, repens.	ROOT perennial, and creeping.
CULMUS in aquofis, bipedalis et ultra, foliofus, nodofus, ftriatus, triqueter, angulis acutis, afperis.	STALK in wet fituations, two feet high, and upwards, leafy, jointed, ftriated, three cornered, the angles fharp, and rough.
FOLIA tres lineas lata, glauca, carinata, & margines carinamque afpera	LEAVES, three lines in breadth, glaucous, keeled, the edges and keel rough.
SPICÆ maiculæ et femineæ diftinctæ, *mafculæ* plerumque tres, erectæ, remotiufculæ, oblongæ, obtufæ, e purpureo-nigræ feu fufcæ, fuprema feffuncioli, inferioribus brevioribus, inæqualibus, bracteatæ, bracteâ inferiore fpicis breviore : *femineæ* duæ, vel tres, longiores, et graciliores, pedunculatæ, plerumque erectæ, apicibus fæpe mafculis.	SPIKES, male and female, diftinct; *male* fpikes generally three, upright, at a little diftance from each other, oblong, obtufe, of a purplifh, black, or brown colour, the uppermoft an inch and a half in length, the lower ones fhorter and unequal, furnifhed with floral leaves, of which the lowermoft is fhorter than the fpikes; *female* fpikes two or three, longer and flanderer than the male, ftanding on footftalks. for the moft part upright. the tips frequently male.
Flos Masc.	*Male Flower.*
SQUAMÆ plurimæ, arcté imbricatæ, ovato-oblongæ, obtufæ, e fufco-purpureæ, nervo medio virefcente. *fig.* 1.	SCALES, numerous, clofely imbricated, of an ovate oblong fhape, obtufe, of a brownifh purple colour, the midrib greenifh, *fig.* 1
STAMINA; FILAMENTA tria, filiformia, alba; ANTHERÆ luteæ. *fig.* 2.	STAMINA: Three FILAMENTS, filiform, and white; ANTHERÆ, yellow, *fig.* 2.
Flos Fem.	*Fem le Flower.*
SQUAMÆ ovato-acuminatæ, fuperne ad lentem denticulis ciliatæ, *fig.* 3.	SCALES ovate, and pointed the upper part when magnified edged with fine teeth, *fig.* 3.
NECTARIUM ovatum, glabrum, ore fiepius bidentatæ. *fig* 4.	NECTARY ovate fmooth. the mouth moft commonly having two teeth, *fig.* 4.
PISTILLUM: GERMEN parvum, intra nectarium; STYLUS nectario paulo longior; STIGMATA tria, patentia. *fig.* 5.	PISTILLUM: GERMEN fmall, within the nectary; STYLE a little longer than the nectary; STIGMATA three, fpreading, *fig.* 5.
SEMEN triquetrum, *fig.* 8. 9. intra nectarium bidentatum. *fig.* 6. 7.	SEED three cornered, *fig.* 8. 9. contained within a nectary having two teeth, *fig.* 6. 7.

the fpikes themfelves, but of the fcales compofing the male fpikes of the *acuta*, and the narrow leaves and flender appearance of the fpikes in the *gracilis*, joined to the want of that glaucous hue in the leaves, fo confpicuous in thofe of the two former; imprefled with thefe general appearances, I carried home their roots, and planted them in my garden, and found at the expiration of two years that they ftill kept up the fame appearances. I then attended more minutely to their parts of fructification, and found fufficient to convince me, and I truft every unprejudiced perfon, that they are three fpecies immutably diftinct.

The largeft and perhaps the moft generally common of the three is our *riparia*, which we have diftinguifhed by that name, from its being found on the edges of rivers, it will alfo grow in the middle of a ditch or pond, and if fuffered to encreafe will quickly fill up any piece of water, being in this refpect almoft equal to the *Poa aquatica*, and *Typha latifolia*; it alfo, by means of its powerfully creeping roots, eafily makes its way through any moorifh ground, and hence is often found in meadows themfelves, and though much fmaller in fuch fituations, its ftriking characters are equally diftinct. Where it grows luxuriantly, its fpikes, efpecially the lowermoft of the female ones, frequently become branched, which gives them a very outré appearance, that may puzzle for a moment: as the male fpikes on their firft appearance are fo eafily diftinguifhed by their pointed and angular appearance, fo the female fpikes, when nearly ripe, are diftinguifhed from the two others by having large, fomewhat inflated, and pointed capfules, flightly bifid at the extremity.

The fynonyms of this and the two other fpecies are fo confounded together, that to trace them through all the writers that have written on the fubject would be an endlefs tafk, it will be fufficient therefore to have quoted a few which may be depended on.

The *Acuta* is next in fize, at leaft with refpect to the breadth of its leaves, to the *Riparia*, and is found in fituations exactly fimilar, indeed they very frequently grow together, and, from the great fimilarity of their foliage, may eafily be confounded; when young, the bluntnefs of its male fpikes and obtufenefs of their Squamæ, fo as totally to want any kind of Arifta, invariably diftinguifhes it from the Riparia, and though there is frequently a tendency in thefe fpikes to be three-cornered, yet the angles are always very obtufe, to which we may add that the colour of them before the Antheræ come forth is much brighter, and fometimes a fpike is found perfectly brilliant; the female fpikes, as well as thofe of the male, are fewer in number, as well as fmaller; nor have they that tendency to be pendulous which thofe of the Riparia frequently have, the Capfules when ripe are alfo much fmaller, more numerous, and no ways inflated, but very fimilar to thofe of the *gracilis*; we may further remark, that while the Squamæ in the male Spikes before the burfting forth of the Antheræ are invariably obtufe, thofe of the female fpikes are pointed, and that while this plant in its ftrong ftate may eafily be miftaken for the Riparia, in its weak ftate it approaches very near the *recurva*, which alfo is a fpecies perfectly diftinct.

Carex acuta

Carex gracilis

CAREX GRACILIS. SLENDER SPIKED CAREX.

CAREX *gracilis* fpicis mafculis et femineis pluribus, fubfiliformibus, floribus digynis.
CAREX nigra verna vulgaris. *Lin. Fl. Lap.* 330. ?
CYPEROIDES angustifolium, caule exquisite triangulari, afpero, fpicis floriferis prælongis, tenuioribus, feminalibus autem fpicis biuncialibus, et habitioribus, erectis, fquamis brevibus acutis, captulis fpadiceo viridibus, rhomboideis, fubtriquetris. *Micheli Nov Gen.* p 60. n. 40.
GRAMEN cyperoides majus angustifolium. *Park,* 1265. *Raii b fl.* 1293. *Syn.* p. 417. n. 2. Great narrow. leaved vernal Cyperus-grafs.

RADIX perennis, repens.

CULMUS in aquofis bi feu tripedalis, in pratis humilior, foliofus, nodofus, triqueter, angulis acutis, afperrimis.

FOLIA *radicalia* longa, viridia, vix glauca, lineas duas lata, ad margines et carinam afpera, vaginantis, *b actralia* lineares cum dimidia lata, inferiore (florente planta) fpicis longiore.

SPICÆ mafculæ et femineæ diftinctæ, mafculæ plerumque tres, e fufco nigricantes, graciles, obfolete triquetræ, nutantes, terminalis biuncialis, inferior duplo aut triplo brevior, infima fæpius androgyna, longior, femineæ tres aut quatuor, teretes, graciles, longitudine mafculi terminalis, fessiles feu breviter pedunculatæ, fuberecti, nigricantes.

MAS.

SQUAMÆ ovato-acutæ, arcte imbricatæ, carinatæ, e purpureo nigricantes, carina, fubviridi. *fig.* 1. auct.

STAMINA: FILAMENTA tria, capillaria, alba; ANTHERÆ lineares, flavæ, *fig.* 2.

FEM.

SQUAMÆ mafc. fimiles, magis vero oblongæ ac obtufæ. *fig.* 3.

NECTARIUM oblongum, glabrum, ore integro; GERMEN minimum; STYLUS nectario longior; STIGMATA duo, villofa, *fig.* 4. 5.

SEMEN triquetrum, minimum, intra nectarium. *fig.* 6.

ROOT perennial and creeping.

STALK, in watery fituations two or three feet high, in meadows not fo tall, leafy, jointed, three cornered, the angles fharp and very rough to the touch.

LEAVES from the *root* long, of a green colour, fcarcely glaucous, two lines in breadth, on the edges and midrib rough, fheathing the ftalk, *b actral* leaves a line and a half in breadth, the lowermoft, while the plant is in flower, longer than the fpikes.

SPIKES, both male and female, growing diftinctly, the *male* generally three in number, of a brownifh black colour, flender, faintly three cornered, drooping, the terminal fpike about two inches in length, the next below twice or thrice as fhort, the lowernmoft for the moft part androgynous and longer, *female* three or four, round, flender, length of the terminal male fpike, feffile or ftanding on fhort footftalks, nearly upright and blackifh.

MALE.

SCALES ovate, pointed, lying clofely one over another, keeled, of purplifh black colour, the keel greenifh, *fig.* 1. *magnif.*

STAMINA: three FILAMENT-, flender and white; ANTHERÆ linear and yellow, *fig.* 2.

FEMALE.

SCALES as in the male, but more oblong and blunter, *fig.* 3.

NECTARY, oblong, fmooth, the mouth entire; GERMEN very fmall; STYLE longer than the Nectary; STIGMATA, two, villous, *fig.* 4. 5.

SEED, three-cornered, very minute, within the nectary, *fig.* 6.

If the feafon be mild, this plant and the Riparia flower in April, and ripen their feeds in June and July.

The *gracilis*, though a flenderer plant both in ftalks, leaves, and fpikes is equal in height where it grows in fimilar fituations to either of the other two, but as this has a greater tendency at leaft in Battersea Meadows to grow among the herbage, it is frequently found fhorter, and fometimes large patches of its foliage are vifible without any flowering fpikes.

This fpecies is diftinguished from the other two, not only by having narrower leaves, which want the glaucous colour of the other two, and flenderer fpikes, which in their young ftate are remarkably pendulous, fo as at firft tight to give this plant an appearance of the *Carex pendula*, but the female flowers are conftantly and invariably digynous. My moft obliging friend Dr. GOODENOUGH, to whom I had communicated my thoughts on this fubject, examining thefe plants with his ufual accuracy, anticipated me in the difcovery of this moft important, moft neceffary character; a character which in a moment decidedly diftinguifhes betwixt two plants, which without it would for ever have been liable to be confounded.

We fhould have been inclined to fuppofe that our *gracilis* was the *acuta* of LINNÆUS, had he not quoted MICHELI's figure, to which he adds the epithet *bona*, that figure is a tolerable reprefentation of our *acuta*, but the fpikes are far too thick for thofe of the *gracilis*.

This fpecies, which is equally common with the two others, flowers a week or two later.

Agriculturally confidered, it is perhaps doubtful, whether we are to rank the Carices with the ufeful or the noxious plants ; from what we have hitherto obferved, we fhould rather clafs them with the latter, not but we think the *Junci, Scirpi,* &c. infinitely more injurious, yet ftill they occupy the room of better grafles ; their principal merit is, that they afford early pafturage, yet their foliage is harfh and rough, and productive of indifferent hay ; and fuch is the opinion of LINNÆUS, who, in his *Flora Lappon,* remarks that the Hufbandman is not fond of fuch meadows as are overrun with Carices, as they afford bad fodder and unprofitable pafturage " nec pinguefcat bos *corice pastus acuta*; unfortunately, however, when the prefent fpecies, or fuch as have fimilar creeping roots, have once got poffeffion of the foil, they are the moft difficult plants poffible to eradicate.

As articles of rural œconomy, they are in many inftances highly ufeful ; in Hampfhire, Surry, and perhaps other hop countries, the leaves of thefe three fpecies are ufed indifcriminately under the name of Sedge, for tying the young hop plants to the poles. MICHELI informs us, that in Italy they are ufed to cover their wine flafks, to make the common fort of chair bottoms, and that the Coopers in making tubs, &c. place them betwixt the ftaves to make them water-tight : to the comfort of the Laplander, they contribute in a high degree by defending him from the feverity

verity of the weather; this is fo particularly defcribed by LINNÆUS in his *Flor. Lappon.* that we fhall tranflate it for fuch of our readers as may not have an opportunity of confulting the original, now become very fcarce.

"Thou wilt wonder, perhaps, curious reader, in what manner human beings are capable of preferving life during "the intenfe feverity of a winter's froft in Lapland, a part of the world deferted on the approach of winter by almoft "every kind of bird and beaft.

"The inhabitants of this inhofpitable climate are obliged to wander with their Rhendeer flocks continually in the "woods; not only in the day-time, but through the longeft winter nights, their cattle are never houfed, nor do they "eat any other food than Liverwort, hence the herdfmen; to fecure them from wild beafts, and other accidents, are "of neceffity kept perpetually with them. The darknefs of their nights is in a great degree overcome and rendered "more tolerable by the light of the ftars reflected from the fnow, and the Aurora Borealis, which in a thoufand fan- "taftic forms nightly illumines their hemifphere. The cold is intenfe, fufficient to frighten and drive us foreigners "from their happy woods. No part of our bodies are fo liable to be deftroyed by cold as the extremities, which "are fituated fartheft from the heart; the chilblains of the hands and feet, fo frequent with us in Sweden, fufficiently "indicate this. In no part of Lapland do we find the inhabitants affected with chilblains, though in refpect to "country one would expect them to be peculiarly fubject to this difeafe, efpecially as they wear no ftockings, while "we cloath ourfelves in one, two, and even three pair.

"A Laplander preferves himfelf from the violence of cold in the following manner; he wears breeches, or rather "trowfers. made of the rough fkin of the Rhendeer, which reach to his ankles, and fhoes made of the fame ma- "terial, the hair turned outward; this grafs, cut down in the fummer, dried, rubbed betwixt the hands, and after- "wards combed or carded, he puts into his fhoes, fo as not only wholly to enwrap his feet. but the lower part of "his legs alfo, which, thus defended, never fuffer from the fevereft cold: with this grafs he alfo fills his hairy "gloves to preferve his hands, and thus are thofe hardy people enabled to bear the froft.

'As this grafs in the winter drives away cold, fo in the fummer it checks the perfpiration of the feet, and pre- "ferves them from being injured by ftones, &c. in travelling, for their fhoes are extremely thin, being made of un- "tanned fkins It is difficult to learn, on enquiry, what the particular fpecies of grafs is which is thus in requeft "with thefe people, as fome ufe one fort, fome another. It is, however, always fome fpecies of Carex, and we "underftood chiefly this."

It is no lefs difficult to underftand what fpecies LINNÆUS himfelf means: he quotes *Morifon's* figure, which is our *fylvat ca*; yet, fays that theCarex grows in *paludibus limo plenis*, which that plant never does with us, it is moft likely, in our opinion, to be one or all of the three common fpecies here figured.

Parietaria officinalis

PARIETARIA *Lin. Gen. Pl.* Polygamia Monoecia.

Hermaphrod. *Cal.* 4-fidus. *Cor.* o. *Stamina* 4. *Styl.* 1. *Sem.* 1. superum, elongatum.

Fem. *Cal.* 4-fidus. *Cor.* o. *Stam.* o. *Stylus* 1. *Sem.* 1. superum, elongatum.

Raii Syn. Gen. 5. Herbae flore imperfecto seu staminzo vel apetalo potius.

PARIETARIA *officinalis* foliis lanceolato-ovatis, pedunculis dichotomis, calycibus diphyllis. *Lin. Syst. Vegetab.* p. 763. *Sp. Pl.* p. 1492.

PARIETARIA foliis elliptico-lanceolatis, hirsutis. *Haller. Hist.* p. 162.

PARIETARIA *officinalis. Scopoli Fl. Carn.* n. 1242.

PARIETARIA officinarum et Dioscoridis. *Bauh. pin.* 121.

HELXINE *Camerar. Epit.* p. 849.

PARIETARIA *Ger. emac.* 331. vulgaris *Parkins.* 437. *Raii Syn.* p. 158. Pellitory of the Wall. *Lightfoot Fl. Scot.* p. 635. *Hudson Fl. Angl. ed.* 2. p. 442. *Order Fl. Dan. I.* 521.

RADIX perennis, sublignosa, rubens, fibrosa.

ROOT perennial, somewhat woody, *of a red colour*, and fibrous.

CAULES plures, suberecti, dodrantales, pedales et ultra, ramosissimi, teretes, striati, solidi, rubentes, pubescentes; *rami* cauli similes, alterni, diffusi.

STALKS several, nearly upright, from nine inches to a foot or more in height, very much branched, round, striated, solid, reddish, and downy; *branches* like the stalks, alternate and spreading.

FOLIA alterna, petiolata, ovata, acuta, utrinque attenuata, integerrima, patentia, ad margines et venas subtus praecipue pubescentia, supra saturate viridia, lucida, subrugosa, punctis prominulis adspersa.

LEAVES alternate, standing on foot-stalks, ovate, pointed, tapering towards each extremity, entire at the edge, spreading, particularly downy at the edge and on the veins of the under-side, on the upper-side of a deep green colour, shining, somewhat wrinkled, and covered over with small prominent points.

PETIOLI longitudine fere diametri foliorum, pubescentes, supra canaliculati.

LEAF-STALKS nearly the length of the diameter of the leaves, downy, hollowed above.

FLORES parvi, herbacei, hirsuti, sessiles, in axillis foliorum conglomerati, hermaphroditi et feminei.

FLOWERS small, of a greenish colour, rough, sessile, growing in clusters in the alae of the leaves, hermaphrodite and female.

Hermaphroditi Flores duo continentur *involucro* heptaphyllo, persistente, foliolis ovatis, acutis planis, hirsutis, hirsutie glandulosa. *fig.* 1.

Two *Hermaphrodite Flowers* are contained in an *involucrum* composed of seven leaves, and permanent, the leaves ovate, pointed, flat, hirsute, the hairs glandular at the extremities. *fig.* 1.

CALYX: Perianthium monophyllum, quadrifidum, planum, persistens. *fig.* 9.

CALYX: a Perianthium of one leaf, deeply divided into four segments, which are flat and permanent. *fig.* 4.

COROLLA nulla, nisi calycem dicas.

COROLLA none, unless the calyx be called so.

STAMINA: Filamenta quatuor, alba, transversim rugosa, instante antheli elastice resilientia, calycemque expandentia; Antherae ovatae, obtusae, didymae; Pollen album. *fig.* 4.

STAMINA: four Filaments of a white colour, wrinkled transversely, on the shedding of the pollen flying back with an elasticity, and expanding the calyx; Antherae ovate, obtuse, double; Pollen white. *fig.* 4.

PISTILLUM: Germen ovatum, viride, nitidum, nudum; Stylus filiformis; Stigma penicilliforme, capitatum, rubertinum. *fig.* 6.

PISTILLUM: Germen ovate, green, shining and naked; Style filiform; Stigma forming a bright scarlet tuft. *fig.* 6.

PERICARPIUM nullum. *Perianthium* elongatum, majus, campanulatum, coloratum, deciduum; ore laciniis conniventibus clauso. *fig.* 3.

SEED-VESSEL none. The *Perianthium* becoming elongated, larger, bell-shaped, coloured and deciduous; the mouth shut by the segments closing together. *fig.* 3.

SEMEN unicum, ovatum, nitidum, in fondo perianthii.

SEED single, ovate, shining in the bottom of the perianthium.

Femineus flos unus inter hermaphroditos ambos, intra involucrum.

One *Female flower* betwixt two hermaphrodite ones, within the involucrum.

CALYX quadrifidus, hirsutus, erectus, germen involvens. *fig.* 5.

CALYX divisible into four segments, hairy, upright, inclosing the germen. *fig.* 5.

COROLLA nulla.

COROLLA none.

PISTILLUM ut hermaphroditi, at stigma majus et paulo inflexum. *fig.* 5.

PISTILLUM as in the hermaphrodites, but the stigma somewhat larger, and bent a little down. *fig.* 5.

PERICARPIUM nullum.

SEED-VESSEL none.

SEMEN unicum ut in hermaphrodito, calyce quadrifido et vix mutato inclusum. *fig.* 7, 8.

SEED single, like that of the hermaphrodites, inclosed in the quadrifid calyx, which is but slightly altered. *fig.* 7, 8.

The

The flowers of the *Parietaria* are so small, and so difficult to investigate, that we need not wonder at their being described differently by different botanists; LINNÆUS's description, in his *Genera Plantarum*, accords best with our observations, his therefore we have adopted with some few alterations.

We find only two sorts of flowers on this plant, viz. hermaphrodite and female; of these, two hermaphrodite and one female blossom are generally placed together in one common involucrum, the female blossom intermediate. To obtain a perfect idea of the manner in which the fructification is carried on, we must examine these flowers at a very early period of their expansion, we shall then find in each involucrum three red stigmata, the two outermost of which belong to hermaphrodite flowers, whose stamina are not yet visible; the middle one, which is largest and most conspicuous, to the female. If we take a view of the same blossoms just at the time that the elastic filaments by their sudden expansion scatter the fertilizing dust of the antheræ, the styles and stigmata of the hermaphrodite flowers, visible before, will often be found wanting, and the germen left naked in the center of the flower; at this period of the blossoming, the segments of the calyx in the same flowers are nearly of the same length as the filaments, the style and stigma of the female blossom remain perfect, with its germen closely surrounded by a green, hairy calyx, which never expands: the blossoming period being now over, a considerable alteration takes place in the calyx of the hermaphrodite flowers, each is considerably elongated, becomes more tubular, assumes a redder colour, has its tips pressed down, and soon drops out of the involucrum, in which it leaves no appearance of a seed; hence I was ready to conclude that these flowers, the imperfection of whose pistilla at a certain age had before been noticed, were certainly barren, but on opening them, I found in the bottom of each a seed perfectly similar to that produced by, and inclosed in the calyx of the female flower, which does not enlarge as the other does, but partaking more of the nature of a capsule, on pressure, divides at top into four parts, and contains a blackish shining seed.

It may seem a little extraordinary, that the imperfect hermaphrodite flowers of this plant should produce perfect seed; but we should consider that they are perfect at first, and that there always is a number of Antheræ belonging to flowers farther advanced bursting near them, from whose pollen they may probably be impregnated.

SCOPOLI describes male flowers on this plant, having a sessile, shining, oblong, and pointed Nectary; surely he must consider the imperfect germen in the hermaphrodite flowers as a Nectarium, otherwise he sees farther than any of his contemporaries.

The curious manner in which these flowers shed their Pollen, or fertilizing dust, is known to most botanists, but may be new to some of our readers; each filament has a peculiarity of structure which renders it highly elastic, there are four of them in number, on their first appearance they all bend inward; as soon as the pollen is arrived at a proper state to be discharged, the warmth of the sun, or the least touch from the point of a pin, will make them instantly fly back with a degree of force, and discharge a little cloud of dust. This process is best seen in a morning, when the sun shines hot on the plant, in July and August; if the plant be large, numbers will be seen exploding at the same instant.

The *Parietaria*, which takes its name from its place of growth, is frequently found on walls, and among rubbish, especially on the walls adjoining the Thames, both above and below Westminster-bridge, it is not a native of *Sweden*, or the more northern countries; this autumn the same degree of cold (viz. about 31 of Fahrenheit's thermometer) which stripped the mulberry of most of its leaves, destroyed the greatest part of its herbage.

Mr. PHILIP MILLER *(vide Dict. ed. 6. 4to.)* asserts that the *Parietaria* which grows wild in England is the Pellitory with a Basil leaf. *Parietaria Ocymi folio* BAUH. *Pin. Parietaria judaica* LIN. and that the *officinalis* LIN. which he says grows naturally in Germany and Holland, was not in England till the year 1727, when he first introduced it; in this opinion Mr. MILLER stands alone, and there is the greatest reason to suppose that he is deceived, and the more so, as the remainder of his account, in which he says that " the seeds are difficult to col-" lect, as they are thrown out of their covers as soon as they are ripe with an elasticity," shows extreme inattention.

As a medicinal plant more virtues appear to have been attributed to the *Parietaria* than it deserves; it has been ranked as an emollient, to which, in the opinion of FLOYER and CULLEN, it has no pretensions, as a diuretic it was an ingredient in the nephritic decoction of the late Edinburgh Dispensatory, which is omitted in the present; in this last intention the expressed juice has been given in the dose of three ounces.

Mr. SOLE, Apothecary of Bath, well known to the Botanic World, for his extensive collection of indigenous plants, informs me that he has observed remarkably good effects from the juice of this herb in dropsical cases, in which other diuretics had failed; he converts the juice into a thin syrup, and gives two table-spoonfuls or more thrice a day.

Monf. TOURNEFORT, speaking of the *Parietaria*, says, " Le sirop de Parietaire soulage fort les hydropiques." *Hist. des Pl. de Paris.* AURELIUS VICTOR informs us, that CONSTANTINE bestowed on the Emperor TRAJAN the name of *Parietaria*, because his statues and his inscriptions, like that herb, were found on all the walls of Rome. *Le Meme.*

It is recommended to be laid on the corn in granaries, for the purpose of driving away that destructive insect the Weevil. *Bradley's Farm. Direct. p. 122.*

EQUISETUM ARVENSE. CORN HORSE-TAIL.

EQUISETUM *Lin. Gen. Pl.* CRYPTOGAMIA FILICES.

    Spica fructificationibus peltatis, bafi dehifcentibus, multivalvi.

Raii Syn. Gen. 4. HERBÆ CAPILLARES ET AFFINES.

EQUISETUM *arvenfe* fcapo fructificante nudo; fterili frondofo. *Lin. Syft. Vegetab.* p. 457. *Sp. Pl.* p. 1516. *Fl. Suec.* n. 928.

EQUISETUM caule florigero nudo, fterili verticillato, radiorum duodecim. *Haller. Hift.* n. 1676.

EQUISETUM *arvenfe. Scopoli Fl. Carn.* n. 1253.

EQUISETUM arvenfe longioribus fetis. *Bauh. Pin.* 16. *Parkins.* 1201. *Raii Hift.* p. 130. Corn Horfe-tail.

EQUISETUM fegetale. *Ger. emac.* 1114.

HIPPURIS minor cum flore. *Dod. Pempt.* p. 73.

EQUISETUM minus terreftre. *L. B. III.* 730. *Hudfon. Fl. Angl. ed.* 2. p. 265. *Lightfoot Fl. Scot.* p. 647.

RADIX perennis, gracilis, nigra, articulata, infigniter reptans, fibris nigricantibus e geniculis exortis capillata.

SCAPI feminiferi ante caules frondofos prodeuntes, et cito marcefcentes, craffitie culmi tritici majoris, palmares aut dodrantales, erecti, nudi, lutefcentes, geniculati, geniculis 2. 3. 5. vaginis multifidis, nervofis, membranaceis, circumveftiti.

SPICÆ feminiferæ terminales, oblongæ, obtufæ, uncinales.

CAPSULÆ feu thecæ feminiferæ plurimæ, angulatæ, erectæ, circa receptaculum proprium collocatæ, et fcuto orbiculato lutefcente tectæ, *fig.* 1. demum introrfum dehifcentes et pulverem virefcentem effundentes, *fig.* 2. 3. auct. *fig.* 4.

CAULIS pedalis et ultra, in apricis obliquus, ftriatos, fcabriufculus, geniculatus. *Vaginæ* numerofæ, denticulatæ.

FOLIA verticillata, octo circiter in fingulo verticillo, fimplices, cauli fimplices.

ROOT perennial, flender, black, jointed, exceedingly creeping, with tufts of black fibres fpringing from the joints.

STALKS producing the feeds fpringing up before the leaf-ftalks, and foon withering, the thicknefs of a large wheat-ftraw, a hand's-breadth or more in height, upright, naked, yellowifh, jointed; joints from two to five, covered with membranous, ribbed fheaths, divided at top into numerous fegments or teeth.

SEED-SPIKES terminal, oblong, obtufe, about an inch in length.

CAPSULES, or feed-cafes, numerous, angular, upright, placed together round a partial receptacle, and covered with a yellowifh orbicular fhield, *fig.* 1. finally opening inwardly, and throwing out a greenifh powder, *fig.* 2. 3. magnified, *fig.* 4.

STALK a foot or more in height, in open fituations oblique, ftriated, roughifh, jointed. The *Sheaths* numerous and toothed.

LEAVES whirled, about eight in each whirl, fimple, and like the ftalk.

Experiment and obfervation, which have difcovered perfect feeds, if not real Stamina and Piftilla, in fome of the plants of this hidden clafs, have hitherto failed in afcertaining what the powder is, which is contained in fuch abundance in the fpikes of the Horfe-tail, different Botanifts differing widely in their opinions concerning it. LINNÆUS and SCOPOLI confider it as the true feed of the plant: HALLER and ADANSON as the male duft; and fo ftrongly was the laft mentioned author of this opinion, that he removed it from the cryptogamous plants, and placed it with the family of the pines, from which, however, as SCOPOLI very judicioufly obferves, it differs *toto cælo.*

We fat down to examine the fructification of this plant, in full expectation of finding Pollen not Seed contained in its cells. We conceived, *a priori,* that a fubftance fo delicately formed, and fo rapid in its growth, could fcarcely produce capfules with ripe feeds; but we rofe from our examination in the full belief of the powders being the real feed, fo far as we could judge from its ftructure and œconomy: actual vegetation muft however be proved, before we can arrive at certainty in this matter.

If a little of the powder be fhaken out of the fpike on a piece of white paper, a moderate magnifier difcovers a motion in it, efpecially if it be breathed on. A fimilar motion is obferved in the capfules of the Ferns when they throw out their feeds; and in the fine powder contained in the heads of the Jungermannia, if we apply a very confiderable magnifier to this powder, we find its motion arife from a very different principle from that which actuated the two former. Here every particle of the powder has three or four, very rarely five, fine, pellucid threads, which are club-fhaped at the extremity. Thefe threads are elaftic, and, by varioufly extending and curling themfelves up, occafion the motion which is fo perceptible. Monf. ADANSON, who has minutely defcribed this feed as pollen, compares the appearance it fometimes affumes to a fpider with its legs ftretched out. Although thefe motions, which are extremely fingular and diverting, are plainly feen with a good magnifier, we never could difcover the body of the feed to make any fort of explofion as Pollen would, under fimilar circumftances of expofure.

The proper time to examine this curious phænomenon is in April, when the plant produces its fpikes.

The medicinal virtues of the Equifetum are too problematical to deferve notice. Writers on the *Materia Medica* rank it with the aftringents.

The Farmer is deeply interefted in a knowledge of this plant, as it is not only one of the moft troublefome and difficult weeds to extirpate that we are acquainted with, but noxious even to cattle, efpecially kine. HALLER relates a particular cafe in which it proved fatal to a young heifer, by bringing on an incurable *Diarrhœa.*

Many parts of *Batterfea Fields* exhibit this plant in perfection. It fometimes is found in meadows and under hedges; and, according to its fituation, like all other plants, affumes a variety of appearances.

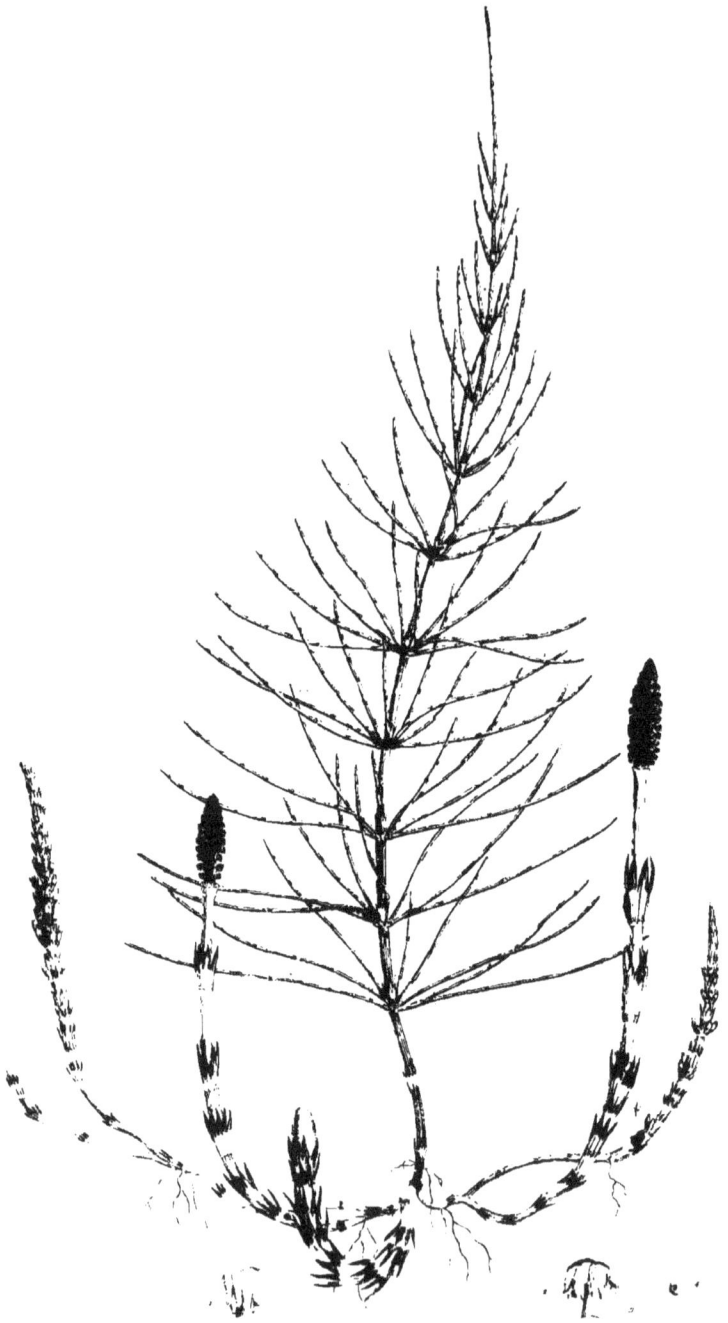

Equisetum arvense

BRYUM BARBATUM. BEARDED BRYUM.

BRYUM *Lin. Gen. Pl.* CRYPTOGAMIA MUSCI.

Anthera operculata. *Calyptra* lævis. *Filamentum* e tuberculo terminali ortum.

Raii Syn. Gen. 1. MUSCI.

BRYUM *barbatum* antheris erectis oblongis barbatis, operculo acuminato obliquo, pedunculis lateralibus.

BRYUM unguiculatum et barbatum tenuius et stellatum. *Dillen. Musc.* tab. 48. fig. 48. Small star-topped, clawed, and bearded Heads.

DENSISSIMIS cæspitibus modo Bryi purpurei ad muros et aggeres nascitur. *fig.* 1.

CAULICULI semunciales aut paulo plus, ramosi, erecti, foliosi. *fig.* 2, 3.

FOLIA e luteo-viridia, densa, lanceolata, acuta, inferioribus suberectis, supremis patentibus. *fig.* 4. auct.

PEDUNCULI semunciales et ultra, ex imo seu ex medio nunquam e summitate surculi enati, singulares vel plures ex eodem surculo, rubentes, nitidi, flexuosi, bulbillo oblongo nudo superne rubro præditi. *fig.* 5.

CAPSULÆ suberectæ, tenues, oblongæ, acuminatæ, olivaceæ, nitidæ. *fig.* 6. *Calyptra* longa, acuminata, parum obliqua. *fig.* 7. *Operculum* longum, tenue. *fig.* 8. *Ciliæ* aurantiacæ, seu coccineæ spiræ modo contortæ. *fig.* 9.

FORMS a thick turf on walls and banks, in the manner of the Bryum purpureum. *fig.* 1.

STALKS half an inch or somewhat more in height, branched, upright, and leafy. *fig.* 2, 3.

LEAVES of a yellowish-green colour, growing thick together, lanceolate, pointed, the lowermost nearly upright, the uppermost spreading 1 *fig.* 4. magnified.

PEDUNCLES half an inch or more in length, proceeding from the bottom or middle, but never from the top of the surculus, one or several from the same surculus, red, shining, crooked, furnished at bottom with a naked oblong bulb, red at top. *fig.* 5.

CAPSULES nearly upright, slender, oblong, pointed, of an olive colour, and shining. *fig.* 6. *Calyptra* long, pointed, a little oblique. *fig.* 7. *Operculum* long, and slender. *fig.* 8. *Ciliæ* orange-coloured or scarlet, twisted in the form of a screw. *fig.* 9.

The moss here represented is undoubtedly the *Bryum* figured by DILLENIUS, in his forty-fifth plate, forty-eighth figure. Neither LINNÆUS nor Mr. HUDSON make mention of it. We are convinced, however, from repeated observations, that it is a species perfectly distinct. It approaches very near to the *Bryum imberbe* and *unguiculatum*. From the former it differs in having the Antheræ or Capsules terminated by long twisted ciliæ, and in having the peduncles always proceeding from the base of the surculus: this last character also strikingly distinguishes it from the *unguiculatum*.

It is not unfrequent about *London*, especially in the environs of *Charlton*, on walls, and barren hilly ground, with the *Bryum purpureum* and *cæspititium*, and produces its fructifications in *December*, *January*, and *February*.

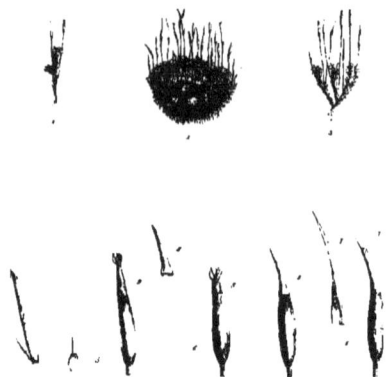

Bryum barbatum.

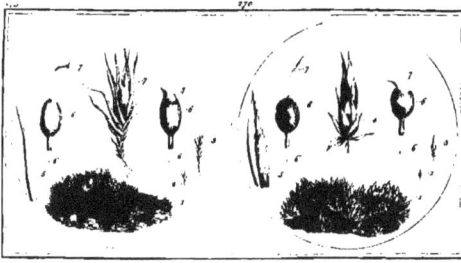

Phascum subulatum.   *Phascum acaulon.*

PHASCUM ACAULON. COMMON PHASCUM.

PHASCUM *Lin. Gen. Pl.* CRYPTOGAMIA MUSCI.
        Calyptra minima. *Operculum nullum.*
    Raii Syn. Gen. 3. MUSCI.
PHASCUM *acaulon* anthera seffili foliis ovatis acutis. *Lin. Syst. Veget.* p. 794. *Sp. Pl.* 1570. *Fl. Suec.* 960.
PHASCUM *cuspidatum* caulescens, foliis ovatis cuspidatis patulis: terminalibus erectis conniventibus. *Schreber. de Phasco.* t. 1. f. 1, 2.
SPHAGNUM acaulon foliis ovato-lanceolatis, pilo aristatis. *Haller. Hist.* 1726.
SPHAGNUM acaulon bulbiforme majus. *Dill. Musc.* 251. t. 32. f. 11. *Raii Syn.* 105. *Lightfoot Fl. Scot.* p. 695. *Hudson Fl. Angl.* p. 466. *Oeder Fl. Dan.* t. 249. i. 1.

PHASCUM SUBULATUM. HEATH PHASCUM.

PHASCUM *subulatum* acaule, anthera seffili, foliis subulato-setaceis patulis. *Lin. Syst. Vegetab.* p. 794. *Sp. Pl.* 1570.
PHASCUM *subulatum* caulescens, foliis lanceolato-linearibus patulis. *Schreber de Phasco*, p. 80.
SPHAGNUM acaulon trichodes. *Haller. Hist.* n. 1727. *Dillen. Hist. Musc.* 251. t. 32. f. 10. *Oeder Fl. Dan.* t. 249. *Lightfoot Fl. Scot.* p. 693. *Hudson. Fl. Angl.* p. 466.

DILLENIUS, who drew the figures of his incomparable work on moffes, without ufing glaffes of any confiderably magnifying power, or, perhaps, without attending fo much to the minutiæ of the parts of fructification as the practice is at prefent, defcribed thefe plants as *having no Calyptra*, and united them with the genus *Sphagnum*. LINNÆUS afterwards made a diftinct genus of them, but without correcting the error of his predeceffor. The following is the defcription of the genus *Phafcum*, which he gives in th 6th edition of the *Genera Plantarum*.
      Mafculus flos fubfeffilis vel brevi pedunculo.
  CAL. Calyptra oblia.
  ANTHERA ovalis ore cibato, tecta *operculo* acuminato.
  RECEPT. *Apophyfis* nulla.
      Femineus flos.
Several Botanifts afterwards defcribing and delineating the *calyptra*, LINNÆUS, in the 13th edition of his *Syftema Naturæ*, publifhed by MURRAY under the title of *Syftema Vegetabilium*, alters it thus:
  ANTHERA operculata: ore ciliata. *Calyptra* minuta, minuta.
This generic defcription, thus altered, is adopted by Mr. HUDSON, without any remarks on its inapplicability to the Englifh *Phafca*. He fhould have informed us, that, however well the character might accord with any foreign Phafcum, the *Englifh* ones, at leaft both thefe here figured, which are the moft common, have neither *Operculum* nor *Ciliæ*. Of this we are fully convinced from repeated examination; and have the pleafure of having our experiments confirmed by the accurate and ingenious SCHREBER, who, in his moft excellent *Monographia, Obfervationes de Phafco*, has the following paffages: " Suturam, qua plerorumque mufcorum vafcula infra apicem, ubi " deinde operculum abfcedit, cinguntur, in nulla hujus generis fpecie obfervare potui. Operculum enim Phafcis " in univerfum omnibus deeft, et vafculum undique in extremum apicem ufque clauditur, fine ullius determinatæ " aperturæ veftigio, quamobrem compreffum utcumque rumpi folet. Cilia igitur in quocumque Phafco fruftra " quærerentur."
  Again, fpeaking of the Capfule, he fays: " *Pollen* in ea nullus adeft nec ullam aperturæ cujufdam determinatæ " aut foraminis, emiffioni contentorum infervientis, veftigium reperire unquam potui. Non enim fponte aperitur, " fed integra perfiftit; donec maxima foliorum pars putruerit, quod in *Ph. pilifero* fæpius obfervavi, aut e foliis " apertis integra elabitur."
  This fingular deficiency then, both of the Operculum and Ciliæ, forms, in our opinion, the beft criterion by which to diftinguifh this genus of plants, and we have altered the generic character accordingly.

PHASCUM ACAULON.

THERE is no mofs more common on the moift banks about *London* than the *Phafcum acaulon*; but as it exhibits no appearance of fructification till in an advanced ftate, and then only to the inquifitive obferver, it generally paffes unnoticed. Mr. LIGHTFOOT gives a fhort, but very expreffive, defcription of it, as follows: " The " leaves, when young, connive together, in " the form of a fmall oval bulb, about one- " fixth of an inch long, and hide the capfule, " which is oval and orange-coloured at firft, but " when ripe fufcous and fhining, and about the " fize of a millet feed." We have reprefented the plant at *fig.* 1. as it ufually grows on the ground; *fig.* 2, 3. detached plants of their natural fize; *fig.* 4. a plant magnified; *fig.* 5. a fingle leaf magnified; *fig.* 6. the anthera or capfule magnified; *fig.* 7. the calyptra alfo magnified.

PHASCUM SUBULATUM.

THE *Phafcum fubulatum* is not fo frequently met with as the *acaulon*, yet is not uncommon on heaths, in the fand pits about *Chorlton*, and on dry banks in a variety of places. They are both found in fructification from *December* to *January*.
In this fpecies the capfule, though fmaller, is much more diftinctly feen, and its calyptra is vifible even to the naked eye. Like the other, it varies much in fize, as alfo in the length, of its peduncle. Mr. LIGHTFOOT confiders it as the leaft of our *Englifh* plants; but Mr. DICKSON, of *Covent-Garden*, who may juftly be called *maximus in minimis*, has difcovered a Phafcum, the *ferratum* of SCHREBER, which is certainly ten times fmaller. *Fig.* 1. reprefents the fubulatum as it ufually grows; *fig.* 2, 3. detached plants; *fig.* 4. a plant magnified; *fig.* 5. a fingle leaf magnified; *fig.* 6. the capfule; *fig.* 7. the calyptra magnified.

JUNGERMANNIA COMPLANATA. FLAT JUNGER-
MANNIA.

JUNGERMANNIA. *Lin. Gen. Pl.* CRYPTOGAMIA ALGÆ.

> MASC. pedunculatus, nudus. *Anthera* quadrivalvis.
> FÆM. feffilis, nudus, feminibus fubrotundis.

Raii Syn. Gen. 18. MUSCI.

JUNGERMANNIA *complanata* furculis repentibus, foliolis inferne auriculatis, duplicato-imbricatis, ramis æqualibus. *Lin. Syft. Vegetab.* p. 803. *Sp. Pl.* p. 1599. *Fl. Suec.* 1041. *Weis. Cryptog.* p. 124.

JUNGERMANNIA foliis rotundis alterne imbricatis, caule plano multifloro, fetis breviffimis. *Haller Hift.* n. 1860.

JUNGERMANNIA complanata: furculo reptante, foliis fubrotundis ferie duplici ordinatis, fubtus appendiculatis; vaginis ramorum plano-truncatis. *Necker. Meth. Mufc.* p. 142.

LICHENASTRUM imbricatum majus fquamis compreffis et planis. *Dillen. Mufc.* 496. *t.* 72. *f.* 26.

JUNGERMANNIA foliis circinatis imbricatim difpofitis ex viridi flavefcentibus. *Michel. Gen.* 7. *t.* 5. *f.* 21.

LICHENASTRUM imbricatum majus. *Raii Syn.* 111. *Hudfon. Fl. Angl.* p. 514. *Lightfoot Fl. Scot.* p. 781.

Defcriptio ex WEIS.	Defcription from WEIS.
SURCULIS variæ longitudinis, ab unciali ad biuncialem longitudinem, planis, inordinate ramofis, ad cortices arborum, latis cæfpitibus repit.	SURCULI of various lengths, from one to two inches, flat, irregularly branched, creeping on the bark of the trees in large patches.
FOLIOLA denfe imbricata, alterna, rotunda, fubpellucida, plana, furculi nervum fuperne tegunt; inferne nervo adhærent parvæ, rotundæ fquamulæ. Color pallide e luteo viridis: Recens mollis tactu eft, aqua madida tota flaccefcit. Ad extremitates, et paffim ad exortum ramolorum prodeunt *thecæ* planæ, fquamis duplo vel triplo longiores, dilute virides, truncatæ, e quibus *fetæ* breves, lineam non excedentes emergunt, tenuiffimæ, cum parvis nigris capitulis, in fufcas lacinulas pilofas diffilientes.	LEAVES of a pale yellow-green colour, clofely imbricated, alternate, round, fomewhat tranfparent, flat, above entirely covering the mid-rib of the furculus, beneath fmall round fcales adhere to the mid-rib, the whole plant, when frefh, is foft to the touch, moiftened with water it grows flaccid. At the extremities, and here and there at the origin of the branches, proceed flat *fheaths* appearing truncated or cut off at top, from whence proceed fhort fetæ or peduncles about a line in length, very flender, and terminated by fmall black heads fplitting into four brown hairy fegments.
Fig. 1. Planta magn. nat.	*Fig.* 1. The plant of its natural fize.
Fig. 2. Pars ejufdem lente aucta.	*Fig.* 2. A part of the fame magnified.
Fig. 3. Pars ejufdem inferior.	*Fig.* 3. The underfide of the fame.
Fig. 4. Theca feu Vagitta.	*Fig.* 4. The Cafe or Sheath.
Fig. 5. Pedunculus.	*Fig.* 5. The Peduncle.
Fig. 6. Capitulum adhuc integrum.	*Fig.* 6. The Capitulum as yet entire.
Fig. 7. Capitulum findens pulveremque fpargens.	*Fig.* 7. The Capitulum fplitting and difcharging its powder.
Fig. 8. Capitulum demiffo pulvere.	*Fig.* 8. The Capitulum with the powder difcharged.

The name of *Jungermannia* was given to this genus by MICHELI in honour of JUNGERMANNUS, a botanift of the laft century.

As LINNÆUS confiders the Capfules of the Moffes as the *Antheræ* containing Pollen, fo in like manner he regards the *Capitula* of the prefent genus as containing Pollen alfo, and the little apparently pulverulent balls, vifible only on a few of the *Jungermanniæ*, but found on moft of the *Muia* as the female flowers, producing feeds. Vid. *Generic Character.*

With all due deference to fuch refpectable authority, we are of opinion, that the character of this genus would be lefs complex, and equally complete, without calling in thofe balls or fphærophylli, as NECKER terms them, at all.

The *Capitula*, or little heads, fplitting when ripe into four parts, replete with a fine powder (whether pollen or feed is immaterial) attached to numerous hairs growing to the infide of the Capitula, are characters which will in all cafes fufficiently diftinguifh this genus. The little heads of thefe plants are capable of affording much microfcopic entertainment. Take a head ready to burft open, place it before the microfcope, affift its opening with the point of a needle, and the elaftic hairs on the infide will inftantly appear in motion, and throw off the globules attached to them in great numbers, and with confiderable force.

The prefent fpecies is one of the moft common of this genus, and may be found in great plenty and perfection about the end of *January*, fpreading on the bark of the Oak and other trees in woods, particularly *Charlton Wood*. It is diftinguifhable from another, equally common, by the pale green colour of its leaves.

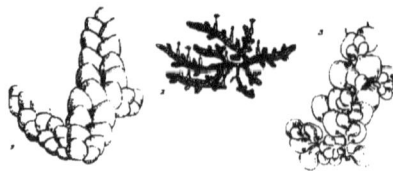

Jungermannia complanata

AGARICUS PROCERUS. TALL MUSHROOM.

AGARICUS *Lin. Gen. Pl.* CRYPTOGAMIA FUNGI.

Fungus horizontalis, fubtus lamellofus.

Raii Syn. Gen. 1. FUNGI.

AGARICUS *anulatus* ftipitatus, pileo campanulato fubfufco fquamofo, lamellis albidis, ftipite bulbofo anulato. *Lightfoot Fl. Scot.* p. 1025.

AGARICUS *procerus* ftipitatus pileo hemifphærico lacerato-fquamofo rufefcente cinereo, lamellis albis, ftipite longo cylindrico acetabulo inferto. *Hudfon Fl. Angl. ed.* 2. p. 612.

AGARICUS *procerus. Scopoli Fl. Carn.* n. 1465.

AMANITA petiolo procero, anulato, in acetabulum pilei immiffo, pileo fquamofo et maculato, lamellis albis. *Haller. Hift.* n. 2371.

AGARICUS; pileo papillari, ampliffimo, in margine fornicato, lacero et filamentofo; petiolo anulato procero, inferius tumido, pilei acetabulo inferto. *Gleditfch. Fung.* p. 114.

FUNGUS pileolo lato, longiffimo pediculo variegato. *C. B. Pin.* 371. n. 24.

FUNGI longiffimo pediculo candicantes, fed maculati efculenti. *J. B.* III. 826. *Raii Syn.* p. 3. n. 10. *Schæff. Fung.* t. 22, 23.

STIPES folitarius, fpithamæus, et ultra, craffitie indicis, bafi bulbofus, fiftulofus, *fig.* 3, albidus, fquamis fufcis plerumque notatus.

STALK fingle, fix inches or more in height, the thicknefs of the forefinger, bulbous at bottom, hollow, *fig.* 3, whitifh, and generally mottled with brown fcales.

VOLVA ampla, perfiftens, lacera, bilamellofa, lamella inferiore membranacea, fubfufca, fuperiore alba, fpongiola.

RUFFLE large, permanent, torn, compofed of two lamellæ or coats, the lowermoft membranous, and brownifh, the uppermoft white and fpongy.

PILEUS palmaris et ultra, primo fubglobofus, dein campanulatus, demum planus, fquamis fufcis, floccidis, circa verticem crebioribus, maculatus; vertex tumidus, coriaceus; caro craffitie lamellarum, alba, molliffima, fpongiofa.

CAP three inches and more in diameter, at firft roundifh, then bell-fhaped, and laftly flat, fpotted with brown, flaky, fcales; thickeft round the crown; crown prominent and leathery, flefh the thicknefs of the gills, white, very foft and fpongy.

LAMELLÆ confertæ, fragiles, albidæ, bafi in marginem acetabuli pilei infertæ, pulverem fubtiliffimum cinerafcentem fpargentes. *fig.* 1, 2.

GILLS numerous, brittle, whitifh, inferted at their bafe into the edge of the cup of the cap, throwing out a very fine afh-coloured powder. *fig.* 1, 2.

This Mufhroom, inferior to few in point of elegance, is frequently found in Woods, and dry hilly Paftures, among Heath, Broom, &c. in the months of September and October; I have found it in the environs of *Hornfey Wood*, and *The Spaniard, Hampftead Heath*; in *Richmond Park*, and many other places.

It is a well-known Mufhroom, and eafily diftinguifhed from all others by its tallnefs, its bulbous bafe, its large ruffle, its fpongy cap, which is flaky, or fcaly, not warty at top, and which, as Mr. LIGHTFOOT juftly obferves, feparates it from the *verrucofus*, with which it has no fmall affinity.

I have feen it expofed to fale in Covent Garden market, for the true eatable one, but a Connoiffeur will diftinguifh it by the fponginefs of its flefh, which renders it in a great degree unfit for eating.

Agaricus Procerus

AGARICUS VELUTIPES. VELVET-STALK'D MUSHROOM.

AGARICUS *Lin. Gen. Pl.* CRYPTOGAMIA FUNGI.

Fungus horizontalis, fubtus lamellofus.

Raii Syn. Gen. 1. FUNGI.

AGARICUS *velutipes* fafciculofus, pileo planiufculo fulvo, ftipite nudo, tenerrime villofo, fuliginofo.

FUNGUS glutinofus colore aurantio. *Vaillant Bot. Paris.* p. 72. 8. t. 12. *fig.* 8, 9.

FUNGUS fafciculofus, pileo orbiculari lutefcente, pediculo fufco, tenerrime villofo, lamellis ex flavo candicantibus, *Raii Syn. ed.* 3. p. 9.

Ex ligno putrefcente, feu arboribus cæfis, coacervatim plerumque nafcitur hic fungus.

VOLVA ad radicem nulla.

STIPES in plantâ mediæ magnitudinis longitudine indicem, craffitie pennam anferinam æquat, teres, haud infrequenter compreffus, fiftulofus, tenerrime villofus, feu velutinus, interne præfertim in vetuftis e rufo-nigricans, fuligine quafi infectus, carneâ citrinâ, in tenuiffima fila fericea partibili.

ANNULUS nullus.

PILEUS uncialis ad triuncialem, modice convexus, fæpe difformis, fulvus, glutinofus ; LAMELLÆ plurimæ, inequales, ex albido-lutefcentes, in majoribus et fenefcentibus fungis craffæ, coriaceæ, lutex, ad marginem pilei undulatæ, longioribus bafi auriculatis, omnibus venofis ; POLLEN cinercum.

This Fungus ufually grows in clufters, out of decayed wood or felled trees.

SHEATH or egg at the root wanting.

STALK in plants of a middle fize about the length of the forefinger, and thicknefs of a goofe-quill, round, not unfrequently flat, hollow, velvety, on the lower part, efpecially in the old ones, of a reddifh black colour, as if tinged with foot, the flefh citron coloured, and feparable into filk-like threads.

RING, or ruffle, wanting.

CAP from one to three inches in diameter, moderately convex, often fhapelefs, of an orange or tawny colour, flimy ; GILLS numerous, unequal, of a whitifh yellow colour, in the larger and older fungi thick, leathery, yellow, waved towards the edge of the cap, the longer ones ear'd at the bafe, and all of them veiny ; POLLEN, or feed, afh-coloured.

Many of the fungi, like the more perfect plants, make their appearance only at a certain period of the year ; others are continually fpringing up and producing their fructifications, almoft regardlefs of particular feafons, of the latter kind is the fungus here figured, at leaft it may be found from September to January in the greateft plenty.

It ufually grows out of dead, or decaying wood, efpecially willow and elm, and almoft always in clufters of different fizes, according to particular fituations ; at the bottoms of old willow-trees I have often feen fifty or more in a clufter, with the caps of fome of the largeft almoft as broad as the palm of one's hand, while on fmall rotten fticks the clufter has confifted of not more than three or four, with caps not more than half an inch in diameter, but however it may vary in regard to fize, &c. it luckily has a character which always readily diftinguifhes it, and this is its velvety and footy ftalk, moft confpicuous in thofe which are advanced.

RAY's fynonym and defcription correfpond pretty accurately with our fungus ; Mr. HUDSON appears to have overlooked or confounded it with fome other fpecies ; VAILLANT has given a good figure, and accurate defcription of it, in his moft elegant work the *Botanicon Parifienfe.*

To the tafte this mufhroom is rather agreeable, and devoid of all acrimony, perhaps it may be eaten with fafety, it is not however of the kind recommended by HORACE,

——— *pratenfibus optima fungis*
Natura eft ; aliis male creditur.

Agaricus Velutipes

AGARICUS FLOCCOSUS. SHAGGY MUSHROOM.

AGARICUS *Lin. Gen. Pl.* FUNGI.

> *Fungus* horizontalis, fubtus lamellofus.

Raii Syn. Gen. 1. FUNGI.

AGARICUS *floccofus* ftipitatus fafciculofus, pileo ftipiteque pilofo-fquamofis, e flavo-fufcis.

PICROMYCES tunicatus. *Batarr.* p. 47. t. 8. H.

AGARICUS floccofus. *Schæffer. Icon. Fung.* t. 61.

Habitat ad radices arborum, plerumque cefpitofus.	Found at the roots of trees, and generally in clufters.
STIPES palmaris et ultra, craffitie digiti minimi feu major, fubcylindraceus, firmus, carnofus, vix fiftulofus, interne albus, fupra annulum nudus, infra filamentofus, pileo concolor.	STALK four inches or more in height, the thicknefs of the little finger or larger, fomewhat cylindrical, firm, flefhy, fcarcely hollow, white within, above the ring naked, below fhaggy, of the fame colour as the cap.
ANNULUS parvus, paulo infra pileum pofitus.	RING fmall, placed a little beneath the cap.
VELUM araneofum, fugaciffimum.	VEIL cob-webby and very fugacious.
PILEUS: *Pilei* diameter fefquinncialis ad palmarem, flavo-fufcus, convexus, in centro nonnunquam faftigiatus, pilofo-fquamofus. *Lamellæ* plurimæ, confertæ, inæquales, ex albo lutefcentes.	CAP: from an inch and a half to four inches in diameter, of a yellow-brown colour, convex, fometimes rifing to a point in the middle, fhaggy. *Gills* numerous, clofe, irregular, of a yellowifh-white colour.

It doth not appear, that this Mufhroom is defcribed either by Mr. RAY or Mr. HUDSON. It approaches very near to the *fquamofus* of the latter, and of which Baron HALLER feems difpofed to confider it as a variety; to us it appears to be a fpecies perfectly diftinct.

BATARRA gives an indifferent figure of it; SCHÆFFER an exceeding good one, very expreffive of the plant we intend: the fpecimens from whence his drawings were made feem to have been fmaller than ours, and fome of them more pointed, which they fometimes are. Thofe reprefented on our plate were found about the middle of *October*, growing at the bottom of a pear-tree in the garden of Mr. *John Chorley*, at *South Lambeth*, where they come up regularly every year. I have alfo frequently found this fpecies in the *Oak of Honour Wood* near *Peckham*. So far as I have remarked, it always grows out of wood, in which refpect it differs from the *fimetarius*, which alfo has a ragged head, but grows out of earth, and has a much longer cap.

When young this Fungus is principally diftinguifhed by the roughnefs of its cap, which appears almoft prickly. Its colour varies from a dingy to a more lively brown.

It is not of the eatable kind; nor do we know any inftance of it proving poifonous: the maggots of flies devour it.

Agaricus floccosus.

Boletus lucidus. Lacquered Boletus.

BOLETUS. *Lin. Gen. Pl.* CRYPTOGAMIA FUNGI. *Fungus* horizontalis fubtus porofus. *Raii Syn. Gen.* 1. FUNGI.

BOLETUS *lucidus*, pileo coriaceo fuperne caftaneo lucido, fulcis circularibus notato, inferne albo ; poris minutiffimis.

BOLETUS *rugofus*. *Jacquin. Fl. Auftr.* t. 169.

FUNGUS coriaceus, pileolo latiffimo atrorubente, pediculo breviffimo. *Raii Syn. ed.* 3. p. 11. ?

FUNGUS fpeciofus putridis arborum truncis et ftipitibus prefertim coryli innafcitur, totus lignofo-coriaceus et perfiftens.

STIPES durus, inæqualis, caftaneus, vernice veluti obductus.

PILEUS plerumque fubdimidiatus, nonnunquam orbiculatus, planus, fuperne caftaneus, nitidiffimus, fulcis circularibus concentricis notatus, rugofus, inferne dum recens albiffimus ; poris minutiffimis.

FUNGI duo connati, in tabula proponuntur, hinc atque illinc fpectati, tum *fungi* pars inferior lente aucta ut pori magis luculenter appareant.

This handfome Fungus grows out of the trunks of decayed trees, particularly the flumps of the hazel, and is throughout of a leathery or fomewhat woody fubftance, and permanent.

STALK hard, uneven, of a chefnut colour, fhining as if varnifhed.

CAP for the moft part forming half a circle, fometimes a whole one, flat, on the upper fide of a chefnut colour, and highly polifhed, marked with circular concentric grooves, the edge thick and wrinkled ; on the under fide, when frefh, very white, the pores exceedingly fmall.

Two Fungi growing together, are reprefented on the plate in two different views ; alfo part of the under fide magnified, to fhew the pores more plainly.

In the month of November, 1780, I fortunately found the fine fpecimen of this Boletus, exhibited on the plate, in the Wood adjoining the Oak of Honour, near Peckham ; on firft difcovering it, the top of the Pileus and ftalk were of fo bright a colour, and fo beautifully polifhed, that I fcarcely knew whether I had found a natural or an artificial production, a view of its under fide, however, foon convinced me it was natural ; it grew out of a rotten hazel ftump.

One principal character of this Fungus, is its polifhed or rather lacquered furface, for it has all the appearance of having been varnifhed ; this I believe it poffeffes in all its fituations ; and on this account I have given it a name different from *Jacquin*, who has an excellent figure of it under the title of *rugofus*, in his *Fl. Auftriaca*. The other characters which feem to be conftant, are the depreffed circles on the upper fide of the Pileus, its wrinkled, thick, and fomewhat inverted edge, which is very apt to break out on the under fide of the Pileus, as reprefented on the plate, but omitted by Jacquin ; the unufual whitenefs of the pileus on its under fide and the extreme finenefs of its pores, which may be overlooked ; it appears to be inconftant in the fhape of its pileus 'which is fometimes *orbiculatus*, as is fhewn on the plate in a fmaller Fungus of the preceding year, and fometimes *dimidiatus*) ; and in its ftalk, which is fometimes wanting, as I difcovered from a fpecimen growing out of the bottom of an elm tree near Hyde Park.

The ftalk of the fpecimen figured, which I preferve, has not fhrunk at all, the pileus is about one third lefs, but retains its form and much of its beauty.

There was fome reafon to fufpect this Boletus of being the *perennis* of LINNÆUS, but the defcription given of that plant in his *Flora Lapponica* removed every doubt, and convinced me he had not defcribed it.

Boletus lucidus.

Phallus caninus.

PHALLUS CANINUS. RED-HEADED MORELL.

PHALLUS *Lin. Gen. Pl.* CRYPTOGAMIA FUNGI.

Fungus supra reticulatus, subtus lævis.

Raii Syn. Gen. FUNGI.

PHALLUS *caninus* volvatus, stipitatus, stipite cellulofo, capitulo impervio, rubro, rugofo.

PHALLUS exilis Marattæ. *Batarr. Fung.* p. 7. t. 40. F. ?

PHALLUS *caninus* volvatus stipitatus, pileo rubro cellulofo acuto, apice claufo. *Hudfon. Fl. Angl. ed.* 2. p. 630.

VOLVA magnitudine nucis mofchatæ, oblongo-ovata, alba, lævis, intus gelatinofa, tunicâ interiore fuperne truncatâ, *fig.* 1, 2.

STIPES extra volvam, fefquiuncialis, feu biuncialis, magnitudine calami anferini majoris, teres, filiformis, inferne acuminatus, cellulofus, fub-pellucidus, pallide aurantiacus, intus cavus, cito flaccefcens. *fig.* 3, 4, 5, 6.

CAPITULUM, nam pileus vix dici poteft, ftipiti inffi-det, eftque feffile, femunciale, diametro fti-pitis, oblongum, fubacuminatum, apice im-pervio albefcente, primo lividum, membranâ nitidâ, tenuiffimâ tectum, infra quam exigua quantitas humoris virefcentis, feu materies fe-minalis fere inodori cernitur, quâ remotâ fu-perficies capituli rubra et tranfverfim rugofa apparet, nequaquam vero cellulofa, ficut in impudico.

VOLVA, or egg, the fize of a nutmeg, of an oblong, ovate fhape; white, fmooth, gelatinous within, the inner coat cut off at top. *fig.* 1, 2.

STALK, beyond the volva, an inch and a half or two inches in length, the fize of a large goofe-quill, round, filiform, terminating in a point at bottom, cellular, fomewhat tranfparent, of a pale orange colour, hollow within, foon be-coming flaccid. *fig.* 3, 4, 5, 6.

HEAD, for it cannot properly be called a cap, fits on the ftem, is feffile, about half an inch in length, and of the diameter of the ftem, ob-long, a little pointed, impervious and whitifh at top, at firft of a livid colour, and covered with a very thin, fhining membrane, under-neath which is a fmall quantity of a greenifh liquid, or feminal matter, almoft fcentlefs, which being removed, the furface of the head appears of a red colour, and tranfverfely wrinkled, but by no means cellular, as in the ftinking Morell.

Mr. EHRET, the celebrated botanic painter, appears to have been the firft who difcovered this rare Fungus in this country; he found it in a wood near *Salop**, and made drawings of it for one of his principal patrons. Mr. HUNTER, gardener to the Earl of MANSFIELD, lately found it, though very fparingly, in *Caen Wood*, and com-municated feveral fpecimens of it to Mr. DICKSON of Covent Garden. This autumn 1781, on the 20th of Sep-tember, I was fortunate enough to be prefent at the difcovering of one of them in Lord MANSFIELD's fmall Pine wood, famous for producing the *Phallus impudicus, Hydnum aurifcalpium,* and other Fungi; I was in fearch of thefe, when my draughtfman Mr. SOWERBY pointed out to me a white fubftance, rifen a little above the furface of the ground, and which at a diftance refembled the cap of a fmall white mufhroom; not fufpecting it to be any thing extraordinary, I took it up with lefs caution than I fhould otherwife have done, and on opening it found it to be the *Phallus caninus,* in the ftate reprefented at *fig.* 1. From the hafty manner in which it was gathered, I had no opportunity of obferving whether its roots were fimilar to thofe of the *Phallus impudicus,* but fufpect they were; on examining it the next morning I was pleafed to find that the ftalk had fhot out from its inclofing volva more than an inch; the volva contained a jelly in the manner of the *impudicus,* in paffing through which, the ftalk became covered with it, (this is prevented from taking place in the *impudicus,* becaufe the pileus is much wider than the ftalk) the ftalk was cellular and hollow, equally rapid in its growth as the *impudicus,* but as its bafe con-tained within the volva ran out to a finer point, fo the body of it was more uniformly of a fize throughout, and of a faint orange colour; not having that firm waxy texture which enables the *impudicus* to fupport itfelf for many days, it quickly became flaccid after attaining its full growth.

Thus far we may obferve a great fimilarity in the ftructure and œconomy of the two plants we have been com-paring; in the remaining part, containing the fructification, we find an amazing difference. The *Phallus impu-dicus* carries on the top of the ftalk a very diftinct and perfect Pileus, or Cap, on the outfide of which the feminal matter is depofited in cells, without the leaft covering; in the *caninus* there is properly fpeaking no Pileus, the part on the outfide of which the feminal matter is lodged, forms a capitulum, or head, which is only a continua-tion of the ftalk, as appears on diffection, differing in its ftructure and colour, this head has a *wrinkled,* not a *reticulated* furface, within thefe wrinkles, which are not very deep, the feminal matter is contained, and (contrary to what we find in the *impudicus)* covered by a very thin membrane; we may obferve that this matter has very little fmell in it, nor do flies appear particularly fond of it.

This unufual ftructure of the capitulum by no means agrees with LINNÆUS's generic character of a *Phallus,* as that implies a *Pileus fmooth on the under, and reticulated on the outer fide,* with which the *impudicus* perfectly cor-refponds, and yet every botanift would call this a *Phallus;* hence there appears a neceffity for altering its generic character, the effence of which feems to confift in the *Seeds being contained in a jelly-like liquifying fubftance,* on the *outfide of a Capitulum or Pileus.*

BATARRA's figure and defcription may poffibly be intended for this Fungus, there is no knowing with cer-tainty, fo great is their obfcurity.

* In the margin of a Ray's Synopfis which had been Mr. Ehret's, belonging to Mr. Richard Haworth, Apothecary of Chancery-lane, who kindly lent it me, there is the following remark at the Phallus impudicus, in Mr. Ehret's own hand writing : " a fmaller fort found in a wood " near Salop, with Mr. Moore 1741, but it did not ftink."

A

C A T A L O G U E

Of certain Plants, growing wild, chiefly in the Environs of SETTLE, in York-shire, obferved by *W. Curtis*, in a Six Weeks Botanical Excurfion from London, made at the Requeft of J. C. LETTSOM, M. D. F. R. S. &c. in the Months of July and Auguft, 1782.

1. *Hippuris vulgaris. Mare's-tail.*
Limnopeuce. *Raii Syn. ed. 2. p.* 136.
In the lakes on *Brig ſtear Moſs*, about four miles from *Kendal* plentifully.

2. *Liguſtrum vulgare. Privet.*
Raii Syn. p. 465.
In *Graſs Wood*, near *Graſſington*, about two miles from *Kilnſay*, not uncommon.

3. *Pinguicula vulgaris. Common Butterwort.*
Pinguicula Geſneri. *Raii Syn. p.* 281.
Common on every Bog.

4. *Utricularia vulgaris. Common-hooded Milfoil.*
Lentibularia. *Raii Syn. p.* 286.
In the greateſt plenty with N° 1. flowers in *Auguſt.*

5. *Schœnus Mariſcus. Long-rooted Baſtard Cyperus.*
Cyperus longus inodorus ſylveſtris. *Raii Syn. p.* 426.
On the edge of *Cowie Tarn*, or Lake, about two miles from *Kendal*, in the greateſt abundance, and higheſt perfection, ſome of the flowering ſtems growing to the height of four or five feet.

6. *Schœnus nigricans. Black Bog-ruſh.*
Juncus lævis minor panicula glomerata nigricante. *Raii Syn. p.* 430.
Plentifully on a Bog in *Skirrith Wood*, near *Ingleton*, and moſt other Bogs in the North.

7. *Schœnus compreſſus. Flat-headed Bog-ruſh.*
Gramen cyperoides ſpica ſimplici compreſſa diſticha. *Raii Syn. p.* 425.
Not uncommon in wet boggy places about *Ingleton*, *Settle*, &c.; near *Gigglefwick* Tarn in plenty; flowers in Auguſt.

8. *Schœnus albus. White Bog-ruſh.*
Cyperus minor paluſtris hirſutus paniculis albis paleacis. *Raii Syn. p.* 427.
On *Brig ſtear Moſs*, n. 1. in abundance.

9. *Scirpus cefpitoſus. Heath Club-ruſh.*
Scirpus montanus capitulo breviori. *Raii Syn. p.* 429.
Frequent on Moors, amongſt the Heath or Ling.

10. *Scirpus acicularis.*
Scirpus minimus capitulis equiſeti. *Raii Syn. p.* 429.
On the edge of a rivulet near *Gigglefwick* Tarn, which runs from the ebbing and flowing Well.

11. *Eriophorum vaginatum. Single-headed Cotton-graſs.*
Juncus alpinus cum cauda leporina. *Raii Syn. p.* 436.
On Peat Bogs frequent, in the afcent to *Ingleborough* Hill.

12. *Melica Montana. Mountain Melic Graſs.*
In *Skirrith Wood*, near *Ingleton*, and in *Graſs Wood*, near *Kilnſay*, not uncommon.

13. *Feſtuca ovina var. vivipara. Viviparous Sheep's Feſcue-graſs.*
Gramen ſparteum montanum ſpica foliacea gramineæ majuſcle minus. *Raii Syn. p.* 410.
On the crags near the ſummit of *Ingleborough*, and and on the rocks of *Long ſtedale*, about ten miles from *Kendal*, plentifully.

14. *Feſtuca elatior. Tall Feſcue-graſs.*
Gramen arundinaceum aquaticum panicula avenacea. *Raii Syn. p.* 511.
On the ſides of the river *Ribble*, near *Settle*, plentifully.

15. *Bromus giganteus. Tall Brome-graſs.*
Gramen avenaceum glabrum, panicula e ſpicis raris ſtrigoſis compoſita, ariſtis tenuiſſimis. *Raii Syn. p.* 415.
Plentifully under the ſtone walls in the road from *Settle* to *Gigglefwick*, and elſewhere.

16. *Bromus hirſutus. Hairy-ſtalked Brome-graſs, Fl. Lond.*
Gramen avenaceum dumetorum panicula ſparſa. *Raii Syn. p.* 51. *nemoralis, Hudſon, Fl. Angl.*
Not uncommon in the woods and hedges in York-shire, eſpecially about *Garrend Wenſleydale*.

17. *Triticum caninum. Bearded Wheat-graſs.*
Gramen caninum ariſtatum radice non repente. *Raii Syn. p.* 58.
Plentiful with the laſt mentioned graſs.

18. *Cynoſurus cæruleus. Blue Dog's-tail graſs.*
Gramen parvum montanum ſpica craſſiore purpureo cœruleo brevi. *Raii Syn. p.* 399.
There is no character in this graſs which would induce one to conſider it as a *Cynoſurus*. Several Botaniſts of the firſt eminence make a diſtinct genus of it, and apparently with much propriety. I firſt found it on the rocks near *Settle*, and afterwards on the rocks and tops of the hills generally in the North. From its being in ſeed when I diſcovered it, which was the latter end of *July*, it muſt be one of the earlieſt graſſes in flower; and of all that I have ever ſeen is by far the moſt hardy. The *Poa pratenſis*, in this reſpect, approaches the neareſt to it, and is very ſimilar in its foliage.
The Botaniſt and the Farmer are both intereſted in the further inveſtigation of this alpine plant.

19. *Scabioſa columbaria. Mountain Scabious.*
Scabioſa minor vulgaris. *Raii Syn. p.* 191.
Frequent on all the lime-ſtone rocks.

20. *Plantago maritima. Sea Plantain.*
Plantago, an alpina anguſtifolia. *Raii Syn. p.* 315.
I found this plant very unexpectedly in the road leading from *Kilnſay* to *Arncliff*, in great abundance; and afterwards diſcovered it on the ſides of many of the mountains thereabouts. I could diſcover no difference betwixt it and the Sea Plantain growing at *Graveſend*.

21. *Sanguiſorba officinalis. Great or Meadow Burnet.*
Sanguiſorba major flore ſpadiceo. *Raii Syn. p.* 203.
Common in moſt of the paſtures; in ſome of which it is the principal plant. The farmers were much divided in their opinions reſpecting its goodneſs. It produces a large, but late crop; grows frequently

quently to the height of four or five feet; but its ſtalks are hard, and apparently unfit for fodder. Some have ſuſpected this was the ſpecies recommended to have been cultivated ſome years ſince; but Dr. WATSON, whoſe authority will not be diſputed, aſſures me, it was the leſſer Burnet, whoſe chief excellence conſiſts in affording foliage early in the ſpring, a property the preſent plant cannot boaſt of.

22. *Aſperula Cynanchica. Squinancy-wort.*
Rubeola vulgaris quadrifolia lævis, floribus purpuraſcentibus. *Raii Syn. p.* 225.
On the lime-ſtone hills about *Conzie,* near *Kendal.*

23. *Galium montanum. Mountain Ladies Bed-ſtraw.*
Mollugo montana minor Gallio albo ſimilis. *Raii Syn. p.* 224.
The moſt general plant on all the Northern mountains. The ſummit of *Ingleborough* is principally covered with it and the *Juncus ſquarroſus.*

24. *Galium boreale. Croſs-leaved Ladies Bed-ſtraw.*
Mollugo montana erecta quadrifolia. *Raii Syn. p.* 221.
Plentiful on the hills betwixt *Kilnſey* and *Graſs Wood,* more eſpecially among ſome rocks in *Graſs Wood.*

25. *Alchemilla vulgaris. Common Ladies Mantle.*
Alchimilla, *Raii Syn. p.* 158.
There is ſcarce a paſture or moiſt bank in Yorkſhire on which this pretty plant does not occur.

26. *Alchemilla alpina. Mountain Ladies Mantle.*
Alchimilla alpina pentaphyllos. *Raii Syn. p.* 158.
This ſpecies, which far excels the other in beauty, is by no means ſo frequent. I firſt found it on a hill called *Lime-ſtone Knot,* in *Loug ſtdale,* a place mentioned by WILSON, in his *Synopſis,* further on in *Long ſtdale,* or, as it is there called, *Long ſteddel,* on the high and romantic rocks about *Buckbarrow Well* the Botaniſt will find it in abundance.

27. *Potamogeton ſetaceum. Setaceous Pond-weed.*
In the ditches on *Brigſtear Moſs,* with the *Utricularia* plentifully.

28. *Anchuſa ſempervirens. Ever-green Alkanet.*
Buglloſſum latifolium ſempervirens. *Raii Syn. p.* 227.
From the paucity of places in which this plant was ſaid to have been found wild, and the ſuſpicion which reſted on ſome of thoſe, I had entertained doubts of its being a native of this country: thoſe were entirely removed on my finding it tolerably plentiful in the lanes about *Gigglſwick,* and in the road between *Sttle* and *Ingleton.*

29. *Primula farinoſa. Birds Eye.*
Verbaſculum umbellatum alpinum minus. *Raii Syn. p.* 285.
Grows every where with the *Pinguicula.* A variety, with white bloſſoms, has not unfrequently been found; in a Bog in *Skirrith Wood,* near *Ingleton,* I gathered ſpecimens a foot and a half high.

30. *Anagallis tenella. Bog Pimpernel.*
Nummularia minor flore purpuraſcente. *Raii Syn. p.* 283.
Common on the Bogs. The largeſt and fineſt plants I ever ſaw of this ſpecies, grew in a bog betwixt *Kendal* and *Long ſtedale.*

31. *Polemonium cæruleum. Jacob's Ladder.*
Polemonium vulgare cœruleum et album. *Raii Syn. p.* 288.
In tolerable plenty at *Malham,* or, as it is there called, *Maum Cove,* by the ſide of the rivulet which ſprings from the baſe of that ſtupendous rock. I found this plant alſo in much greater plenty in *Coniſtone Dib,* in a low wood, under ſhelter of ſome high and romantic rocks, a ſituation it ſeems to affect. Alſo on *Amber Scar,* on the left-hand between *Kilnſey* and *Arncliff.*

32. *Samolus valerandi. Water Pimpernel.*
Samolus valerandi. *Raii Syn.* 283.
Obſerved a few ſpecimens on *Brigſtear Moſs.*

33. *Campanula latifolia. Giant Bell-flower.*
Campanula maxima foliis latiſſimis. *Raii Syn. p.* 276.
Extremely common about *Settle* and elſewhere, under the ſtone-walls and hedges. The country people improperly call them Fox-gloves.

34. *Ribes rubrum. Common Currants.*
Ribes vulgaris flore rubro. *Raii Syn. p.* 456.
On the edges of the wet ditches, and in the Woods about *Carr End, Wenſlydale.*

35. *Gentiana amarella. Autumnal Gentian.*
Gentianella pratenſis flore lanuginoſo. *Raii Syn. p.* 275.
Common on lime-ſtone hills and paſtures.

36. *Ulmus campeſtris* var. *γ Hudſon. Wych Elm.*
Ulmus folio latiſſimo ſcabro. *Raii Syn. p.* 469.
Common in Hedges and Woods. It is much ſuperior in ſize and beauty to the common Elm, and deſerves to be more generally cultivated.

37. *Oenanthe crocata. Hemlock-water Drop-wort.*
Oenanthe cicutæ facie lobelii. *Raii Syn. p.* 210.
In the wet ditches betwixt *Kendal* and *Long ſtedale,* and in ſimilar ſituations, in many parts of Yorkſhire.

38. *Scandix odorata. Sweet Cicely.*
Cerefolium magnum five Myrrhis. *Ger. emac. p.* 1037.
This plant is not mentioned by Mr. RAY in his *Synopſis;* and Mr. HUDSON introduces it as a doubtful native. The ſituations in which I found it clearly convinced me, it had a good title to be conſidered as a native of Great Britain; and I was confirmed in my opinion by the common people, who find it wild in many places in the greateſt plenty, and call it by the name of *Sweet-ciſs,* an abbreviation of *Cicely.* They rub their furniture with it, to give it a gloſs. It is common under the ſtone-walls about *Settle,* at the entrance into *Kilnſey,* and in *Whitfell Gill,* near *Aſtrig.*

39. *Pimpinella Saxifraga. Burnet Saxifrage.*
Pimpinella ſaxifraga minor, foliis ſanguiſorbæ. *Raii Syn. p.* 213.
Very frequent in the fiſſures of the Lime-ſtone Rocks about *Settle.*

40. *Parnaſſia paluſtris. Graſs of Parnaſſus.*
Parnaſſia vulgaris et paluſtris. *Raii Syn. p.* 555.
Very common in Bogs and wet Meadows.

41. *Droſera rotundifolia. Round-leaved Sun-dew.*
Ros ſolis folio rotundo. *Raii Syn. p.* 356.

42. *Droſera longifolia. Long-leaved Sun-dew.*
Ros ſolis folio oblongo. *Raii Syn. p.* 356.
I found theſe two ſpecies plentifully in the North; but no where in greater plenty, or perfection, than on *Brigſtear Moſs,* near *Kendal,* where they grow to twice or thrice the ſize they uſually acquire with us; but in other reſpects appeared to me to afford no truly ſpecific character. It is very probable, that the three ſpecies enumerated in RAY, in addition to the above, will be found to be varieties only.

43. *Allium arenarium? Sand Garlic.*
Allium ſylveſtre amphicarpon foliis porraceis, floribus et nucleis purpureis. *Raii Syn. p.* 370.

44. *Allium oleraceum? Herbaceous Garlic.*
Allium ſylveſtre bicorne flore ex herbaceo albicante cum triplici in ſingulis petalis ſtria atro-purpurea. *Raii Syn. p.* 370.
Theſe two ſpecies of Garlic being out of flower when diſcovered, I dare not be poſitive about them. The firſt grew ſparingly, in a paſture at the back of the ſtables belonging to the *Dun Horſe, Ingleton,* kept by Mr. *Wariner,* at whoſe houſe every traveller finds himſelf at home. The latter grew alſo ſparingly among rocks, in the *Girbing Trough,* near *Coniſtone, Kilnſew.*

45. *Anthericum oſſifragum. Lancaſhire Aſphodel.*
Phalangium anglicum paluſtre Iridis folio. *Raii Syn. p.* 375.

Extremely

Extremely common in all Bogs and moorish Grounds, which in *July* and *Augu?* are beautifully decorated with its bloſſoms.

46. *Convallaria Polygonatum. Sweet Solomon's Seal.*
Polygonatum floribus ex ſingularibus pediculis. *Raii Syn. p. 263.*
In the rocky part of *Sykes Wood*, near *Ingleton*, ſparingly.

47. *Juncus ſylvaticus. Great hairy Wood Ruſh.*
Gramen nemoroſum hirſutum latifolium maximum. *Raii Syn. p. 416.*
In *Whitſell Gell*, near *Aſkrig*, plentifully: alſo near the bottom of a mountain called the *Rye-loaf*, near *Settle*, where no wood was growing, but probably had grown.

48. *Triglochin paluſtre. Arrow-headed graſs.*
Common in marſhy places.

49. *Rumex digynus. Mountain Sorrel.*
Acetoſa rotundifolia repens Eboracenſis, folio in medio deliquium patiente. *Raii Syn. p. 143.*
Found ſparingly in the ſpot mentioned by RAY, cloſe by *Buckbarrow Well*, in *Longſtedale*, on the edge of a deep rivulet abounding in waterfalls. The *Rumex ſcutatus* of LINNÆUS is very common in the gardens in *Yorkſhire*: I have ſometimes ſeen it in ſituations which have tempted me to think it an indigenous plant.

50. *Colchicum autumnale. Meadow Saffron.*
Colchicum commune. *Raii Syn. p. 373.*
Not uncommon in the meadows in *Yorkſhire*, I found it in a paſture cloſe by *Milſcur Luſh*, near *Kilnſay*, Mr. WM. FOTHERGILL, of *Carr End*, informed me, that it grew plentifully in a meadow near *Weſt Witton*, *Wenſleydale*.

51. *Aliſma ranunculoides. Small Water Plantain.*
Plantago aquatica minor. *Raii Syn. p. 357.*
In *Gigglſwick* Tarn plentifully.

52. *Epilobium anguſtifolium. Roſe bay Willow-herb.*
Lyſimachia ſpecioſa quibuſdam Onagra dicta ſiliquoſa. *Raii Syn. p. 310.*
In *Graſs Wood*, near *Kilnſay*, among the rocks, plentifully in one particular ſpot.

53. *Epilobium alpinum. Alpine Willow-herb.*
Lyſimachia ſiliquoſa glabra minor latifolia. *Raii Syn. p. 311.*
On the moiſt rocks about *Buckbarrow Well*.

54. *Vaccinium Myrtillus. Blea-berry.*
Vitis idæa anguloſa. *Raii Syn. p. 457.*
Common on all the Heaths, Rocks, and Mountains.

55. *Vaccinium Vitis idæa. Red Bil-berry.*
Vitis idæa ſempervirens fructu rubro. *Raii Syn. p. 457.*
Not uncommon on Heaths, yet ſeldom found in bloſſom.

56. *Vaccinium Oxycoccos. Cran-berry.*
Oxycoccos ſ. Vaccinia paluſtria. *Raii Syn. p. 267.*
Frequent on the boggy moſſes about *Settle*, *Kendal*, and elſewhere in the North.

57. *Polygonum viviparum. Viviparous Biſtort.*
Biſtorta minor. *Raii Syn. p. 147.*
On the edge of *Semer Water*, an extenſive tarn at *Carr End*, *Wenſleydale.*

58. *Paris quadrifolia. Herb Paris, or True-love.*
Herba Paris. *Raii Syn. p. 264.*
In *Kelkoe Wood*, near *Settle*, and moſt of the woods thereabout.

59. *Pyrola rotundifolia. Common Winter-green.*
In the enchanting woods of *Heckfall*, near *Grewelthorpe*, in tolerable plenty. Sparingly in *Roydale Wood*, near *Carr End*, *Wenſleydale*; alſo in *Tennants Wood*, near *Kilnſay.*

60. *Saxifraga ſtellaris. Hoary Kidney-wort.*
Geum paluſtre minus foliis oblongis crenatis. *Raii Syn. p. 354.*
Not uncommon on the moiſt rocks and boggy ground about *Buckbarrow Well*, *Longſtedale*; a few plants in bloſſom, but moſtly in ſeed.

61. *Saxifraga oppoſitifolia. Purple Saxifrage.*
Saxifraga alpina ericoides, flore cæruleo. *Raii Syn. p. 353.*

On the craggy rocks of *Ingleborough* and *Pennigent* plentifully, in particular ſpots.

62. *Saxifraga autumnalis. Autumnal Saxifrage.*
Saxifraga alpina anguſtifolia, flore luteo guttato. *Raii Syn. p. 353.*
On the moiſt rocks of *Ingleborough* ſparingly. In the greateſt plenty in *Longſtedale*; alſo in *Whitſell Gill*, near *Aſkrig*, moſt beautifully in bloſſom.

63. *Saxifraga hypnoides. Trifid Saxifrage — Ladies Cuſhion.*
Saxifraga muſcoſa trifido folio. *Raii Syn. p. 354.*
On the mountains about *Settle* plentifully, and moſt of the mountains in the North.

64. *Arenaria verna. Mountain Sandwort or Chickweed.*
Alſine puſilla pulchro flore folio tenuiſſimo noſtras. *Raii Syn. p. 351.*
Generally with the laſt mentioned plant. I always found it a ſure indication of elevated ground.

65. *Sedum anglicum. Engliſh Stonecrop.*
Sedum minimum non acre flore albo. *Raii Syn. p. 271.*
On ſome rocks in *Longſtedale*, on the left-hand ſide going down the vale; obſerved it on a few rocks only.

66. *Sedum villoſum. Hairy Stonecrop.*
Sedum purpureum prateuſe. *Raii Syn. p. 270.*
On the ſide of *Ingleborough* where the ſprings originate; but in much greater plenty in ſimilar ſituations about *Carr End*, *Wenſleydale.*

67. *Spergula nodoſa. Knotted Spurrey.*
Alſine paluſtris foliis tenuiſſimis, ſeu Saxifraga paluſtris anglica. *Raii Syn. p. 350.*
Common on the Bogs about *Settle*, and ſimilar ſituations in the North.

68. *Prunus Padus. Bird Cherry.*
Ceraſus avium nigra et racemoſa. *Raii Syn. p. 463.*
In the woods about *Ingleborough*, and elſewhere in the North, plentifully.

69. *Cratægus Aria. White Beam Tree.*
Meſpilus alni folio ſubtus incano, Aria Theophraſti dicta. *Raii Syn. p. 453.*
Common in the mountainous woods in the North; loves a dry ſituation.

70. *Roſa villoſa. Apple Roſe.*
Roſa ſylveſtris pomifera major noſtras. *Raii Syn. p. 454.*
In *Graſs Wood*, near *Kilnſay*, and in ſeveral other woods.

71. *Rubus idæus. Raſberry.*
Rubus Idæus ſpinoſus fructu rubro. *Raii Syn. p. 467.*
Plentiful in the above mentioned wood.

72. *Rubus ſaxatilis. Stone Bramble.*
Not unfrequent in the mountainous woods about *Settle* and *Ingleton*; but no where in greater perfection than near the ſummit of *Helſeinab*, near *Kendal.*

73. *Rubus Chamæmorus. Cloud-berry.*
Chamæmorus. *Raii Syn. p. 260.*
On the ſides of the higheſt mountains about *Settle* and *Ingleton*, eſpecially on the *Rye-loaf*, within a few miles of the former, where I gathered its berries in the greateſt perfection, and found the caterpillar of the *Emperor Moth (Phalæna pavonia)* feeding on its foliage.

74. *Potentilla verna. Spring Cinquefoil.*
Pentaphyllum parvum hirſutum. *Raii Syn. p. 255.*
My very obliging friend Mr. WM. FOTHERGILL, of *Carr End*, ſhewed me this plant growing ſparingly on an old ſtone-wall at *Carlow-nick*, adjoining the weſt-end of the *Crag Paſture*, about half a mile from *Carr End*. I have the beſt authority for believing, that the *Potentilla opaca* of Mr. HUDSON is no other than this plant.

75. *Geum rivale. Water Avens.*
Caryophyllata montana purpurea. *Raii Syn. p. 253.*
In the Paſtures, Woods, &c. about *Settle* and elſewhere much more common than the *urbanum* is with us.

7

76.

76. *Dryas octopetala*. *Mountain Dryas*.
Caryophyllata alpina chamaedryos folio. *Raii Syn.*
p. 253.
This beautiful plant, heretofore known to be only
a native of *Scotland* and *Ireland*, I found plentifully in feed on *Arncliff Clouder*, a mountain
within half a mile of *Arncliff*, in *Littendale*, a
few miles from *Kilnsay*.

77. *Comarum palustre*. *Marsh-cinquefoil*.
Pentaphylloides paluftre rubrum. *Raii Syn. p.*
256.
In *Gigglefwick* Tarn, near *Settle*, plentifully, and
other marfhy places.

78. *Actæa spicata*. *Herb Chriftopher, or Bane-berry*.
Chriftophoriana. *Raii Syn.* 262.
I am indebted to Mr. WM. FOTHERGILL, before-
mentioned, for pointing out to me a moft delightful herborizing fpot, viz. a Glen or Gill, called
Whitfell Gill, or *Arthur Fofs*, fituated within a
fmall diftance of *Afkrig*. In this fheltered valley,
ornamented with an enchanting water-fall, many
rare plants grew in the utmoft luxuriance. Here
I found, in abundance, this poifonous plant lurk-
ing, and half concealing its dark glofly berries,
not unaptly refembling thofe of coffee, but more
beautiful, and within reach of my arm, around
one plant of it, the following, viz. *Scandix odo-*
rata, Saxifraga autumnalis, Ribes rubrum, Rubus
idæus, Rubus faxatilis, Prunus Padus, Juncus
fylvaticus. What a treat for a Botanift! What a
recompence for one of the rougheft journies ever
I am, perhaps, ever experienced! Auguft 16.

79. *Aquilegia vulgaris*. *Common Columbine*.
Aquilegia. *Raii Syn. p.* 273.
Found among fome lime-ftones on the upper part of
the *Girling Trough*, near *Coniftone, Kilnsay*, out
of bloom. It poffibly might be the *alpina*.

80. *Thalictrum minus*. *Leffer Meadow-Rue*.
Thalictrum minus. *Raii Syn. p.* 203.
In *Skirrith Wood*, near *Ingleton*, fparingly. In great
plenty on the mountainous ground about *Kilnsay*
and many other places in the north.

81. *Trollius europæus*. *Globe-flower, Locker-gowlons*.
Ranunculus globofus. *Raii Syn. p.* 272.
In *Skirrith Wood*, and the moift woods about *Settle*,
in great abundance.

83. *Galeopfis tetrabit* var. 3. *Nettle Hemp*.
Lamium cannabino folio, flore amplo luteo; labio
purpureo. *Raii Syn. p.* 241.
This elegant variety is found fparingly in the Corn-
fields about *Settle*.

83. *Draba muralis*. *Speed-well-leaved Whitlow-grafs*.
Burfa paftoris major loculo oblongo. *Raii Syn. p.*
292.
On *Arnber Scar*, near *Arncliff*, in *Littendale*, and at
Malbam Cove, fparingly.

84. *Draba incana*. *Wreathen-podded Whitlow-grafs*.
Lunaria contorta major. *Raii Syn. p.* 291.
Very common on the rocks about *Settle*, and fimilar
fituations elfewhere.

85. *Thlafpi montanum*. *Mountain Thlafpi*.
Thlafpi foliis globulariæ. *Raii Syn. p.* 305.
On the mountainous paftures in the road from *Settle*
to *Malbam*, within about half a mile of the Tarn,
plentifully, with the *Arenaria verna*, moftly in
feed. I fought for it in vain in the paftures about
the ebbing and flowing well.

86. *Cochlearia officinalis*. *Common Scurvy-grafs*.
Cochlearia. *Raii Syn. p.* 302.
Common by the river *Ribble*, near *Settle*, and on the
mountains thereabout. In the latter fituation it
is very dwarfifh, and is the *grœnlandica* of LIN-
NÆUS.

87. *Turritis hirfuta*. *Hairy Tower Muftard*.
Turritis muralis minor. *Raii Syn. p.* 294.
On old caftles, walls, and rocks, about *Settle* and
Ingleton, common.

88. *Cardamine impatiens*. *Impatient Ladies-Smock*.
Cardamine impatiens, vulgo ficum minus impatiens.
R i Syn. p. 299.

On *Gigglefwick Scar* fparingly, in feed.

89. *Geranium fylvaticum*. *Wood Cranefbill*.
Geranium batrachoides montanum noftras. *R..*
Syn. p. 360.
In the woods and paftures about *Settle* and *Ingleton*
not uncommon; alfo in *Longftedale*; and about
Carr End, Mr. W. FOTHERGILL informs me, it
is fo common as to empurple the paftures when
in full bloom.

90. *Geranium fanguineum*. *Bloody Cranefbill*.
Geranium hæmatodes. *Raii Syn. p.* 360.
In rocky mountainous woods very common, as in
Kelkoe Wood, near *Settle*, in *Grafs Wood*, and in
the road from thence to *Kilnsay*, in the greateft
plenty.

91. *Geranium columbinum*. *Long-ftalked Cranefbill*.
Geranium columbinum, diffectis foliis, pediculis
florum longiffimis. *Raii Syn. p.* 359.

92. *Geranium lucidum*. *Shining Cranefbill*.
Geranium faxatile. *Raii Syn p.* 361.
Common on the ftone-fences about *Settle* and elfe-
where.

93. *Fumaria claviculata*. *Climbing Fumitory*.
Fumaria alba latifolia. *Raii Syn. p.* 335.
Plentifully on a thatched Farm-houfe in *Longftedale*,
on the left-hand fide, going down the vale.

94. *Vicia fylvatica*. *Wood Vetch*.
Vicia fylvatica multiflora. *Raii Syn. p.* 322.
I found one root only of this beautiful plant in full
bloffom in *Skirrith Wood*, near *Ingleton*.

95. *Hippocrepis comofa*. *Horfe-fhoe Vetch*.
Ferrum equinum germanicum filiquis in fummitate.
Raii Syn. p. 321.
Grows in abundance out of the lime-ftone rocks,
near *Gigglefwick, Settle*, and *Kilnsay*.

96. *Trifolium alpeftre*. *Long-leaved Clover*.
Trifolium purpureum majus, foliis longioribus et
anguftioribus, floribus faturatioribus. *Raii Syn.*
p. 328.
In *Skirrith* and other mountainous woods and paftures in the North, moft plentifully.

97. *Hypericum montanum*. *Mountain St. John's Wort*.
Hypericum elegantiffimum non ramofum folio lato.
Raii Syn. p. 343.
In *Syke's Wood*, near *Ingleton*, and other moun-
tainous woods, not uncommon.

98. *Hieracium murorum*. *Wall Hawkweed, or Golden*
Lung-wort.
Hieracium murorum folio pilofiffimo. *Raii Syn.*
p. 168.
On the rocks near the water-fall at *Afgarth Force*,
and, if I miftake not, on *Kilnsay Crag*.

99. *Hieracium fabaudum*. *Shrubby Hawkweed*.
Hieracium fruticofum latifolium hirfutum. *Raii*
Syn. p. 167.
This plant, in its ufual ftate is extremely common.
A variety, whofe leaves are fpotted with red, and
which is fometimes miftaken for the *Hypochæris*
maculata, is frequent on the rocks in *Grafs Wood*
and at *Gordel*.

100. *Carduus helenioides*. *Melancholy Thiftle*.
Cirfium britanicum Clufii repens. *Raii Syn. p.*
193.
In a coppice near *Gigglefwick* and in *Skirrith Wood*,
plentifully. In the paftures about *Bordley*, near
Malbam, fo plentiful as to empurple the paftures, fo ftriking in its foliage, and fo noxious
in its effects, as to attract the notice of the
hufbandmen, who call them *White-backs*.

101. *Viola paluftris*. *Marfh Violet*.
Viola paluftris rotundifolia glabra. *Raii Syn. p.*
364.
In *Gigglefwick* Tarn, and other marfhy fituations,
common.

102. *Viola grandiflora*. *Yellow Panfie*.
Viola montana lutea grandiflora noftras. *Raii Syn.*
p. 356.
In mountainous paftures frequent, about *Attamire*
Cliffs, near *Settle*.

103.

103. *Orchis bifolia. Butterfly Orchis.*
Orchis alba bifolia minor cultare oblongo. *Raii Syn.* 380.
In the hilly paſtures above *Stackhouſe* and on *Mill Iſland*, near *Settle*, plentifully, and in many other paſtures, fully blown.

104. *Orchis conopſea. Sweet Orchis.*
Orchis palmata rubella cum longis calcaribus rubellis. *Raii Syn.* 380.
On *Mill Iſland* and moſt of the paſtures with the former, fully blown.

105. *Satyrium viride. Frog Orchis.*
Orchis palmata minor flore luteo-viridi. *Raii Syn.* 381.
Frequent on the moſt hilly paſtures about *Settle*, in full bloom.

106. *Ophrys muſcifera. Fly Orchis.*
Orchis myodes galea et alis herbidis. *Raii Syn.* 399.
On the hilly lime-ſtone paſtures at *Stackhouſe*, near *Settle*, plentifully; in *Skirrith Wood*, ſparingly. Mr. ROBERT KIDD, of *Aſhton* near *Gargrave*, ſhewed me one he had gathered in a wild ſtate, two feet and a quarter high, with fourteen bloſſoms on it. He alſo ſhewed me a great number of the *Ophrys apifera*, or *Bee orchis*, a rare plant with them; but whoſe place of growth he did not care to divulge.

107. *Cypripedium Calceolus. Ladies ſlipper.*
Calceolus mariæ. *Raii Syn.* 385.
The beauty and extreme ſingularity of the bloſſoms of this plant, joined to its great ſcarcity, have occaſioned it to be univerſally ſought after by Botaniſts and others; who, not content with contemplating its beauties in its native ſoil, are anxious to ſee it grow in their gardens, in which, however, they are generally diſappointed, as it very rarely thrives on tranſplanting. We ſaw, indeed, a few inſtances to the contrary in ſome gardens in *Yorkſhire*. To this rage for the Ladies Slipper we may attribute its preſent ſcarcity in *Helk's Wood* near *Ingleton*, where it uſed to be found in plenty. We were fortunate enough to diſcover this plant in conſiderable plenty in the neighbourhood of *Kilnſay*, not only in the Woods with its uſual attendant, the red-flowered Helleborine, but alſo in hilly paſture ground, with the *Ophrys ovata*; but as ſome gardeners in the neighbourhood had diſcovered them, and were unremittingly employed in digging up every one they found, we may venture to prophecy, that in a few years they will be rarely found here alſo.

108. *Serapias paluſtris.*
Helleborine paluſtris noſtras. *Raii Syn.* 384.
In the boggy part of *Syke's Wood* plentifully; alſo near *Kilnſay*, and many other boggy ſituations. To us it appears to be a very diſtinct ſpecies. Flowers in July.

109. *Serapias purpuraſcens.*
Helleborine altera atro-rubente flore. *Raii Syn.* 383.
This ſpecies is found in *Syke's Wood*, and is common to moſt of the woods in the North, eſpecially ſuch as are mountainous and rocky; it produces a long ſpike of red or purpliſh flowers, the beginning of Auguſt. This ſpecies is frequently miſtaken for the Ladies Slipper.

110. *Sparganium natans. Small Burr-reed.*
Sparganium non ramoſum. *Raii Syn.* 437. 2. 3.
In the lakes on *Brigſtear Moſs*, with the *Hippuris* and *Utricularia*, not uncommon.

111. *Carex pulicaris. Flea Carex.*
Gramen cyperöides mhimum, ſeminibus deorſum reflexis puliciformibus. *Raii Syn.* 24.
On the ſides of *Ingleborough* and other mountainous ſituations tolerably frequent.

112. *Carex diſtans. Looſe Carex.*
Gramen cyperöides ſpicis parvis longiſſime diſtantibus. *Raii Syn.* 421.

This moſt variable Carex we found in almoſt every ſituation, on the edge of *Gigglefwick Tarn* it grew with the *panicea* plentifully. I alſo found it on dry ground near the tops of the higheſt mountains. Some ſpecimens, in particular, a yard in height, I gathered near the ſummit of a lofty rock in *Langſledale*.

113. *Carex veſicaria. Bladder Carex.*
Gramen cyperöides polyſtachion majus, ſpicis teretibus, erectis. *Raii Syn.* 419.
We do not recollect finding this ſpecies nearer *London* than *Virginia Water*. In the North it is a common Carex on the edges of tarns and rivulets. It abounds in *Gigglefwick Tarn*, a ſpot fertile in Carices, and on the borders of *Semer Water*, *Wenſledale*.

114. *Carex gracilis*, Fl. Lond. *Slender-ſpiked Carex.*
Gramen cyperöides majus anguſtifolium. *Raii Syn.* 417.
In great plenty on the borders of *Conwic Tarn* near *Kendal*.

115. *Salix Pentandra. Sweet Willow.*
Salix folio laureo, ſeu lato glabro odorato. *Raii Syn.* 449.
About *Kilnſay*, and more eſpecially about *Carr End Wenſledale*, this is the moſt common ſpecies of Willow, and is much uſed for making the larger ſort of baſkets. Its leaves are gloſſy, and exhale an odoriferous perfume in hot weather, which, joined to the beautiful appearance of the male-tree when in bloom; and the female when in ſeed, render it one of the moſt deſirable trees our iſland naturally produces.

116. *Salix helix. Spurge-leaved Willow.*
Salix humilior, foliis anguſtis ſubcrenulcis ex adverſo binis. *Raii Syn.* 448.
Equally common with the foregoing, and uſed for making the finer ſorts of baſket-work.

117. *Salix roſmarinifolia.*
We have no doubt but the Willow, to which we aſſign this name, is a ſpecies perfectly diſtinct. It approaches neareſt to the *vitellina*. Its twigs are remarkably tough. We found it on the edge of a rivulet which runs into *Semer Water*, *Wenſledale*. As cuttings of this and the two following Willows, introduced into our garden, have grown, we hope to be able to ſpeak more deciſively on them at ſome future period.

118. *Salix myrſinites?*
One ſmall ſhrub of this ſpecies, which correſponds with the deſcription Mr. LIGHTFOOT gives of the *Myrſinites*, we found with ſeveral of the following on the ſlope of a high hill betwixt *Kilnſay* and *Arncliff*.

119. *Salix arenaria.*

120. *Empetrum nigrum.*
Empetrum montanum fructu nigro. *Raii Syn.* 444.
On the ſides of *Ingleborough* plentifully.

121. *Taxus baccata. Yew-tree.*
Taxus. *Raii Syn.* 445.
Growing in a truly wild ſtate out of the clefts of the rocks on *Gigglefwick Scar*. Dr. ABRAHAM SUTCLIFFE, of *Settle*, to whoſe kind hoſpitality and uſeful information I am much indebted, was an eye-witneſs to the fatal effects of this plant on two Bullocks, who had careleſly been ſuffered to feed on its foliage.

122. *Acer Pſeudoplatanus. Sycamore Maple.*
Acer majus. *Raii Syn.* 470.
Very common in woods, hedges, and round gentleman's ſeats, the latter from its quick growth, its great ſize, and power of reſiſting the moſt violent ſtorms without injury, it is admirably calculated to preſerve. Its wood, though ſeldom uſed in building, is applied to many œconomical purpoſes.

123. *Ophiogloſſum vulgatum. Adders-tongue.*
Ophiogloſſum. *Raii Syn.* 128.

In meadows and by the sides of rivulets much more frequent than with us.

124. *Ofmunda Lunaria. Moon-wort.*
Lunaria minor. *Raii Syn.* 128.
On *Mear Bank* by *Sykes' Wood, Ingleton,* and other places, with the frog Orchis, not unfrequent.

125. *Ofmunda crifpa. Stone-fern.*
Adiantum album crifpum alpinum. *Raii Syn.* 126.
Among the ftones about *Buckburrow Well* in *Lang-fledale,* in the utmoft abundance, and here and there on the walls betwixt that fpot and *Kendal.*

126. *Afplenium Scolopendrium. Harts-tongue.*
Phyllitis. *Raii Syn.* 116.
Between the fiffures of the rocks on the tops of moft of the high mountains.

127. *Afplenium Ruta muraria.*
Ruta muraria. *Raii Syn.* 122.
On the rocks about *Settle,* and elfewhere, very common.

128. *Afplenium Trichomanes. Common Maiden-hair.*
Trichomanes. *Raii Syn.* 119.
Very common on the rocks and ftone fences.

129. *Polypodium Phegopteris. Wood Polypody.*

Filix minor Britannica pediculo pallidiore, alis inferioribus deorfum fpectantibus. *Raii Syn.* 122.
We found one plant of this rare fpecies among our dried fpecimens, but do not recollect its place of growth; fufpect we took it for the following, with which it has fome fimilarity in its general appearance.

130. *Polypodium Dryopteris. Branched Polypody.*
Filix ramofa minor. *Raii Syn.* 125.
We obferved this fpecies in tolerable plenty about *Kinfay,* particularly among loofe lime-ftones on the right-hand fide of the *Girling Trough* near *Coniston.*

131. *Polypodium fragile. Brittle Polypody.*
Filix faxatilis caule tenui ftagile. *Raii Syn.* 125.
Extremely common on old caftles, ftone fences, &c. about *Settle* and elfewhere.

132. *Lycopodium Selago. Fir Club-mofs.*
Selago foliis et facie abietis. *Raii Syn.* 106.

133. *Lycopodium alpinum. Mountain Club-mofs.*
Lycopodium Sabinæ facie. *Raii Syn.* 108.
Both of thefe fpecies are found in abundance near the fummit of *Ingleborough.*

In the courfe of our excurfions we could not avoid noticing, *en paffant,* an almoft infinite number of *Moffes, Lichens,* &c. which particularly abound in moft of the fpots we vifited; but as few of them were in fructification; and as the larger plants were altogether fufficient to engrofs our attention, we muft defer gratifying the curious Cryptogamift till an opportunity prefents itfelf of revifiting thefe delightful regions at a different period of the year.

We may remark, that the *Allium,* which we fuppofed to be the *oleraceum,* proved, on flowering, to be the *carinatum;* and that the *Potentilla,* which has not yet flowered, feems, from its foliage, as if it would prove either a fingular variety of the *verna,* or a diftinct fpecies.

www.ingramcontent.com/pod-product-compliance
Lightning Source LLC
Chambersburg PA
CBHW021513210326
41599CB00012B/1238